[Florida alligator]

BIG CYPRESS NATIONAL PRESERVE, FLORIDA

Located on Florida's southwest coast, Big Cypress National Preserve was created in 1974 to protect the water and other natural resources of the Big Cypress Swamp. The preserve covers an area of 291,375 hectares (720,000 acres).

The fresh waters of the Big Cypress Swamp are important to the health of the surrounding Everglades. The swamp supports a mixture of plants and a variety of wildlife, including the Florida panther, migrating birds, and the Florida alligator.

One of the best times to see alligators is from November through April – the dry season – when large groups of alligators move to deep ponds and canals.

NATIONAL GEOGRAPHIC
SCIENCE

FLORIDA

SCIENCE

NATIONAL GEOGRAPHIC
School Publishing

PROGRAM AUTHORS

Randy Bell, Ph.D.

Malcolm B. Butler, Ph.D.

Kathy Cabe Trundle, Ph.D.

Judith S. Lederman, Ph.D.

David W. Moore, Ph.D.

Program Authors

RANDY BELL, PH.D.

Associate Professor of Science Education,
University of Virginia, Charlottesville, Virginia
SCIENCE

MALCOLM B. BUTLER, PH.D.

Associate Professor of Science Education,
University of South Florida, St. Petersburg, Florida
SCIENCE

KATHY CABE TRUNDLE, PH.D.

Associate Professor of Early Childhood Science
Education, The School of Teaching and Learning,
The Ohio State University, Columbus, Ohio
SCIENCE

JUDITH SWEENEY LEDERMAN, PH.D.

Director of Teacher Education, Associate
Professor of Science Education, Department of
Mathematics and Science Education,
Illinois Institute of Technology, Chicago, Illinois
SCIENCE

DAVID W. MOORE, PH.D.

Professor of Education,
College of Teacher Education and Leadership,
Arizona State University, Tempe, Arizona
LITERACY

Program Reviewers

Miranda Carpenter
Teacher, MS Academy Leader
Imagine School
Bradenton, FL

Kelly Culbert
K–5 Science Lab Teacher
Princeton Elementary
Orange County, FL

Richard Ellenburg
Science Lab Teacher
Camelot Elementary
Orlando, FL

Beth Faulkner
Brevard Public Schools
Elementary Training Cadre,
Science Point of Contact,
Teacher, NBCT
Apollo Elementary
Titusville, FL

Kathleen Jordan
Teacher
Wolf Lake Elementary
Orlando, FL

Melissa Mishovsky
Science Lab Teacher
Palmetto Elementary
Orlando, FL

Shelley Reinacher
Science Coach
Auburndale Central Elementary
Auburndale, FL

Flavia Reyes
Teacher
Oak Hammock Elementary
Port St. Lucie, FL

Rose Sedely
Science Teacher
Eustis Heights Elementary
Eustis, FL

Michelle Thrift
Science Instructor
Durrance Elementary
Orlando, FL

Cathy Trent
Teacher
Ft. Myers Beach Elementary
Ft. Myers Beach, FL

Acknowledgments
Grateful acknowledgment is given to the authors, artists, photographers, museums, publishers, and agents for permission to reprint copyrighted material. Every effort has been made to secure the appropriate permission. If any omissions have been made or if corrections are required, please contact the Publisher.

Illustrator Credits
All illustrations by Precision Graphics. All maps by Mapping Specialists.

Photographic Credits
Front Cover Image Quest Marine/ Alamy Images.

Credits continue on page EM14.

The National Geographic Society
John M. Fahey, Jr.,
President & Chief Executive Officer

Gilbert M. Grosvenor,
Chairman of the Board

Copyright © 2011 The Hampton-Brown Company, Inc., a wholly owned subsidiary of the National Geographic Society, publishing under the imprints National Geographic School Publishing and Hampton-Brown.

National Geographic School Publishing
Hampton-Brown
www.myNGconnect.com

Printed in the USA.
RR Donnelley
Jefferson City, MO

ISBN: 978-0-7362-7732-7

11 12 13 14 15 16 17 18 19 20

2 3 4 5 6 7 8 9 10

LIFE SCIENCE

CONTENTS

TECHTREK
myNGconnect.com

Student eEdition

Vocabulary Games

Digital Library

Enrichment Activities

CHAPTER

3

EARTH SCIENCE

CONTENTS

TECHTREK
myNGconnect.com

 Student eEdition

 Vocabulary Games

 Digital Library

 Enrichment Activities

PHYSICAL SCIENCE

CONTENTS

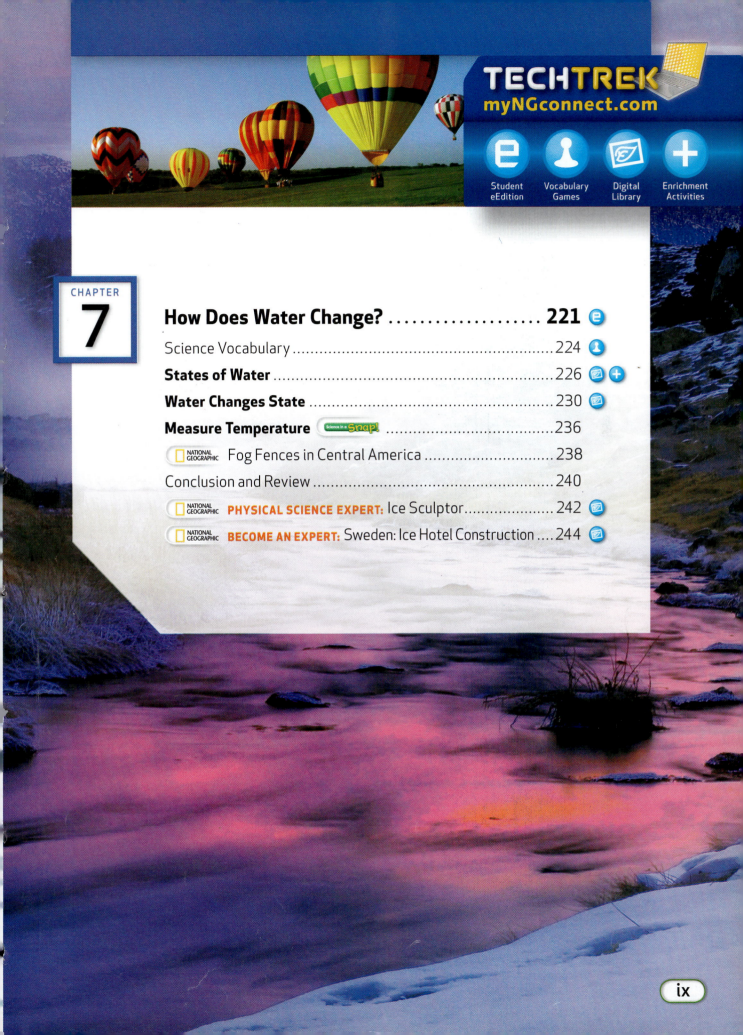

TECHTREK
myNGconnect.com

Student eEdition
Vocabulary Games
Digital Library
Enrichment Activities

CHAPTER
7

TECHTREK
myNGconnect.com

 Student eEdition
 Vocabulary Games
 Digital Library
 Enrichment Activities

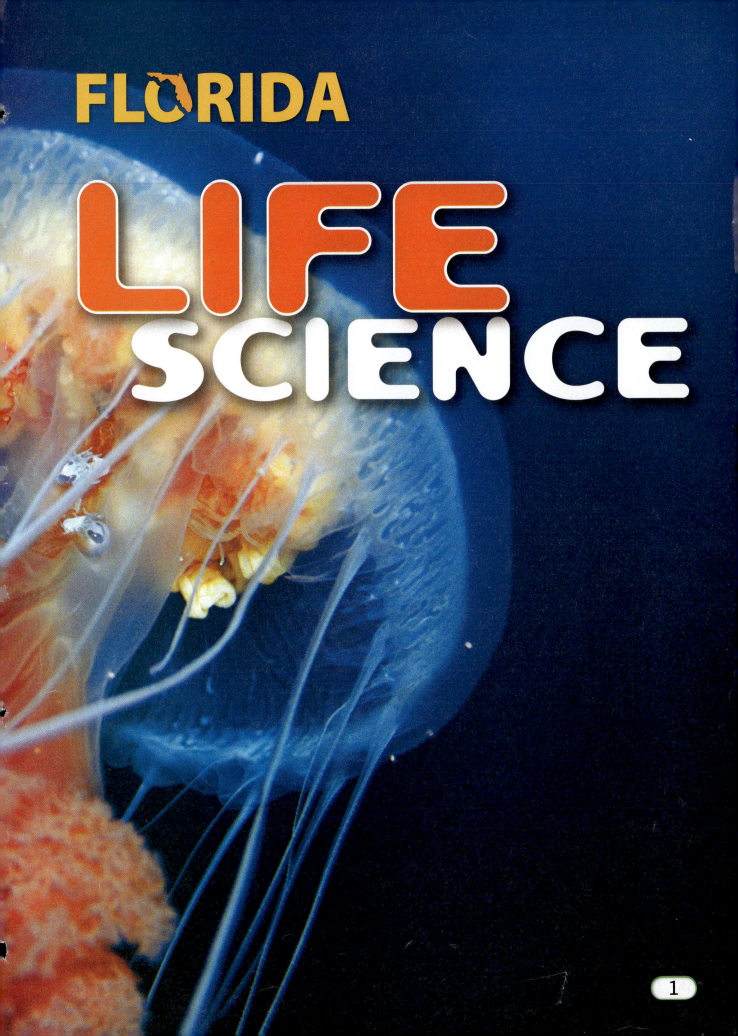

FLORIDA

LIFE SCIENCE

What Is Life Science?

Life science is the study of all the living things around you and how they interact with one another and with the environment. This type of science investigates how living things are similar to and different from one another, how they live and reproduce, and how they function in the environment. Life science includes the study of humans, as well as all the other kinds of living things on Earth. People who study living things and the environment are called life scientists.

You will learn about these aspects of life science in this unit:

HOW DO PLANTS LIVE AND GROW?

Many different kinds of plants exist. Some have flowers, but not all of them do. A flower is just one kind of plant part that carries out a specific job as the plant grows. Life scientists study how plants live and grow.

HOW ARE ANIMALS ALIKE AND DIFFERENT?

Some kinds of animals have backbones while other kinds of animals do not. Animals can be grouped according to characteristics such as backbones. Life scientists study characteristics of animals and group them according to similarities and differences.

HOW DO PLANTS AND ANIMALS RESPOND TO SEASONS?

The seasons bring changes in light, temperature, and rainfall in most environments. Plants and animals respond to these seasonal changes in many different ways. Life scientists study how plants and animals respond to changes in the environment.

MEET A SCIENTIST

Tierney Thys: Marine Biologist, Filmmaker, Pilot

Tierney Thys is a marine biologist, filmmaker, pilot, and National Geographic Emerging Explorer. Since 2000, Thys and her colleagues have been traveling the world's oceans to study the giant ocean sunfish, or mola. Though these fish can grow more than three meters (ten feet) long and weigh over 2,270 kilograms (5,000 pounds), little is known about them. By placing high-tech satellite tags on molas and collecting mola tissue samples for genetic analysis, Thys and her colleagues hope to uncover the mola's secrets.

"When it comes to fishes, the mola really pushes the boundary of fish form," says Tierney. "It seems a somewhat counterintuitive design for swimming in the waters of the open seas—a rather goofy design—and yet the more I learn about it, the more respect and admiration I have for it. That's what makes my work so exciting!"

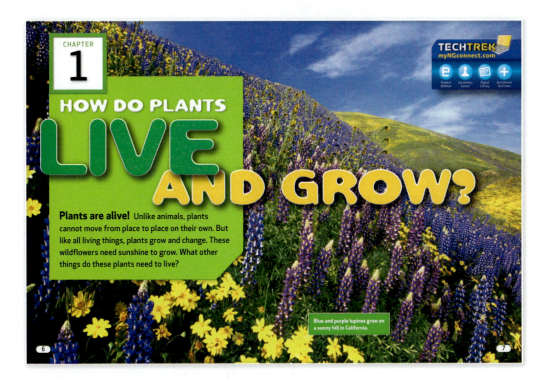

6

CHAPTER
1

HOW DO PLANTS LIVE AND GROW?

TECHTREK
myNGconnect.com

Student eEdition | Vocabulary Games | Digital Library | Enrichment Activities

Plants are alive! Unlike animals, plants cannot move from place to place on their own. But like all living things, plants grow and change. These wildflowers need sunshine to grow. What other things do these plants need to live?

Blue and purple lupines grow on a sunny hill in California.

7

In Chapter 1, you will learn:

FLORIDA NEXT GENERATION SUNSHINE STATE STANDARDS

SC.3.L.14.1 Describe structures in plants and their roles in food production, support, water and nutrient transport, and reproduction. **LEAVES, ROOTS AND STEMS, GROUPS OF PLANTS**

SC.3.L.14.2 Investigate and describe how plants respond to stimuli (heat, light, gravity), such as the way plant stems grow toward light and their roots grow downward in response to gravity. **HOW PLANTS RESPOND**

SC.3.L.15.2 Classify flowering and nonflowering plants into major groups, such as those that produce seeds, or those like ferns and mosses that produce spores, according to their physical characteristics. **GROUPS OF PLANTS**

SC.3.L.17.2 Recognize that plants use energy from the Sun, air, and water to make their own food. **LEAVES**

SC.3.L.14.1 Science in a Snap! Describe structures in plants and their roles in food production, support, water and nutrient transport, and reproduction.

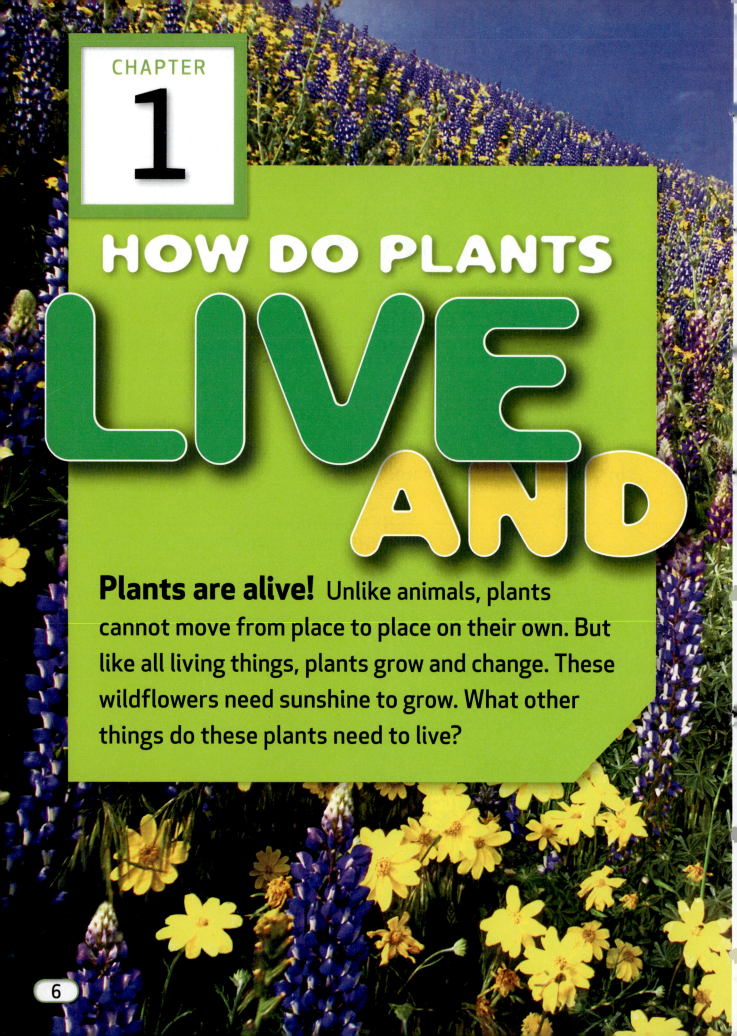

HOW DO PLANTS LIVE AND

Plants are alive! Unlike animals, plants cannot move from place to place on their own. But like all living things, plants grow and change. These wildflowers need sunshine to grow. What other things do these plants need to live?

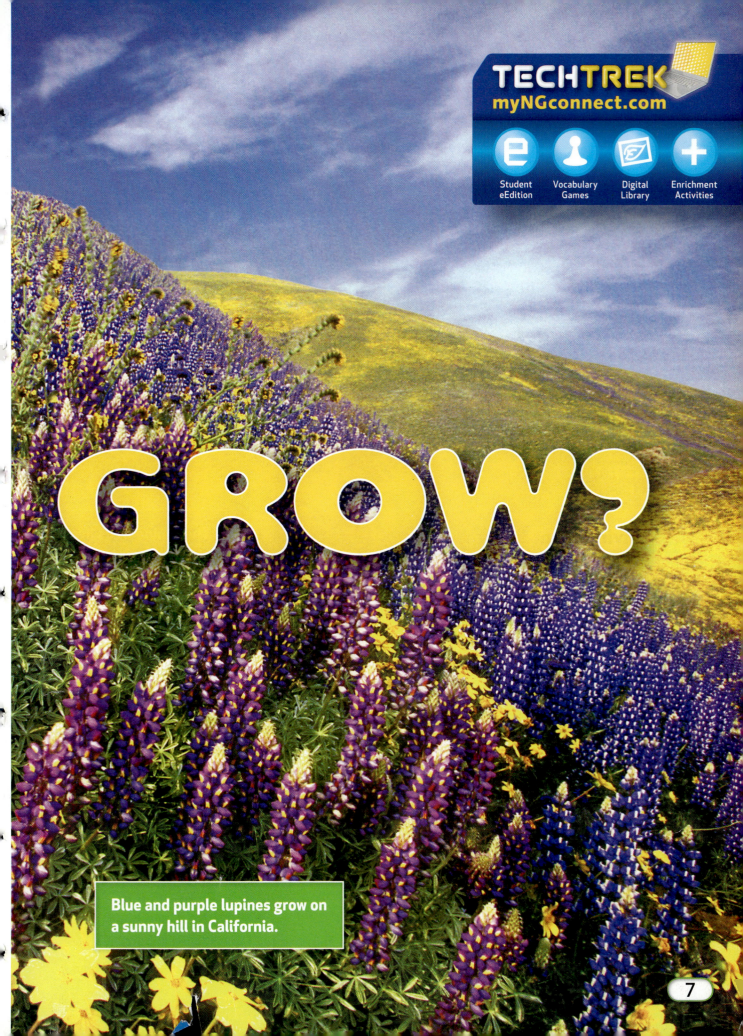

GROW?

Blue and purple lupines grow on
a sunny hill in California.

SCIENCE VOCABULARY

organism (OR-guh-niz-uhm)

An **organism** is a living thing. (p. 10)

A fern is an organism.

environment (en-VI-ruhn-ment)

The **environment** is all the living and nonliving things around an organism. (p. 20)

These pine trees live in a windy environment.

germinate (JUR-muh-NĀT)

Seeds **germinate** when they begin to grow. (p. 21)

A bean seed can germinate if the soil is moist and warm.

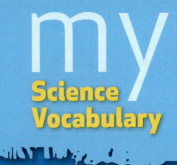

my
Science Vocabulary

environment
(en-VI-ruhn-ment)

germinate
(JUR-muh-NĀT)

organism
(OR-guh-niz-uhm)

pollen
(POL-uhn)

reproduce
(rē-pru-DUS)

spore
(SPOR)

TECHTREK
myNGconnect.com

Vocabulary Games

reproduce (rē-pru-DUS)

To **reproduce** is to make more of its own kind. (p. 26)

Apple trees reproduce by making seeds.

pollen (POL-uhn)

Pollen is a powder made by a male cone or the male parts of a flower. (p. 26)

When bees visit flowers, pollen sticks to their bodies.

spore (SPOR)

A **spore** is a tiny part of a fern or moss that can grow into a new plant. (p. 30)

The spores of a fern grow in cases often found underneath its leaves.

Leaves

Plants are living things, or **organisms**. Like all organisms, plants need water, food for energy, and space to live and grow. Plants also need air and nutrients from the soil.

Most plants have three main parts—leaves, roots, and stems. These parts work together to keep the plant alive.

Leaves use the energy of sunlight to make food for the plant.

How Plants Make Food Almost all plants make their own food. This is usually the job of the leaves. To make food, leaves need sunlight, air, and water.

Leaves capture the energy of sunlight. They use this energy to change air and water into food. Air enters the plant through tiny holes in its leaves. Water comes from the soil. Water travels to the leaves through the roots and stems.

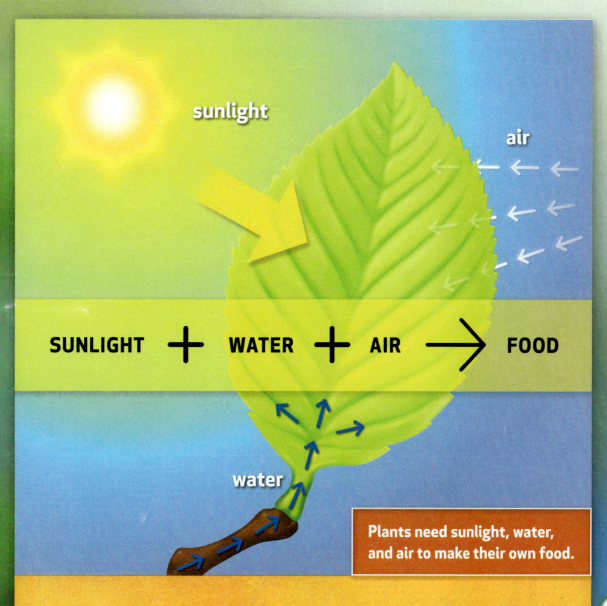

sunlight

air

SUNLIGHT **+** WATER **+** AIR → FOOD

water

Plants need sunlight, water, and air to make their own food.

Kinds of Leaves

Each kind of plant has its own type of leaf. Compare the leaves in the picture below. Some of the leaves are big. Others are divided into many parts.

All leaves have veins that carry water through the leaf. Veins also carry food from the leaves to the rest of the plant. In some plants, the veins branch out in many directions. In other plants, the veins are straight and do not cross each other.

The leaves of flowering plants come in many shapes and sizes.

Digital Library

Most leaves are green. Different kinds of plants have leaves of very different shapes and sizes.

FLOWERING PLANTS
Many flowering plants have leaves that are broad and flat.

PLANTS WITH CONES
Pine trees and many other plants with cones have long, narrow leaves that are shaped like needles.

FERNS
The leaves of ferns are called fronds. Fronds are often divided into many parts.

MOSSES
The green parts of mosses are not true leaves because they do not have veins.

Before You Move On

1. What things do plants need to live and grow?
2. How do leaves make food for a plant?
3. **Analyze** Could the leaves of a plant live without roots and stems? Why?

Roots and Stems

Roots Which part of a plant is usually hidden from sight? The roots! Roots grow down into the soil. The main job of the roots is to take in water and nutrients. Nutrients are parts of soil that plants need to grow and stay healthy. Roots also hold the plant in place and help it stand upright.

The many small roots of grass help hold the soil in place.

Look at the roots in the picture below. Different plants have different kinds of roots. Grasses have many small roots. Other plants have one main root that grows deep into the soil. These long roots can reach water far below the surface. Water moves up the root to the plant's stems and leaves.

This sunflower has one main root that is long and thick.

main root

Stems

Stems support the plant. They hold up the leaves and flowers. The trunk of a tree is a thick, woody stem. The picture shows the long trunks of palm trees. Notice how their trunks hold their leaves high above the ground. Other kinds of stems are slender, such as the stem of a daisy. The stalks of corn are also stems. Can you think of more kinds of stems?

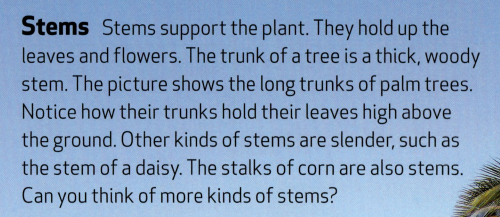

Enrichment Activities

TECHTREK
myNGconnect.com

Stems carry water and food between the roots and leaves.

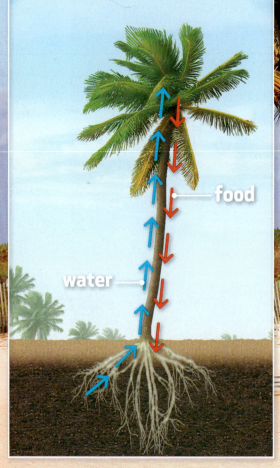

food

water

The trunks of these palm trees are stems.

Stems connect the roots and leaves. Stems carry water and nutrients from the roots to the leaves. Food made in the leaves moves through the stems to the rest of the plant.

Where Does the Water Go?

Fill a cup about half full with water. Add about 20 drops of food coloring.

Your teacher will cut off the bottom of your flower stem. Place the stem in the cup. Wait one day. Then observe the flower.

What do you see? What happened to the colored water?

Before You Move On

1. What does a root take from the soil?
2. How do the roots of a plant get food? List the parts of a plant that food passes through as it travels to the roots.
3. **Generalize** How are the stalk of a bean plant and the trunk of a tree alike? How do a stalk and a trunk help a plant to live?

NATIONAL GEOGRAPHIC

POTATOES: ROOTS OR STEMS?

Do you like potatoes? Millions of people do!

Potatoes were first grown by people who lived in the Andes Mountains of South America. About 500 years ago, Spanish explorers brought potatoes to Europe. From Europe, farmers brought potatoes to many other parts of the world.

These farmers in Peru are harvesting potatoes.

Potatoes are still an important food for people in the Andes. They grow potatoes of many different shapes and sizes. Their potatoes also come in many different colors—green, yellow, red, and even purple!

Potatoes grow in the ground, but they are not roots. Potatoes are stems that store food for the plant. The leaves of a potato plant use sunlight to make food. Some of this food is then stored underground. When farmers harvest potatoes, you get to eat that food!

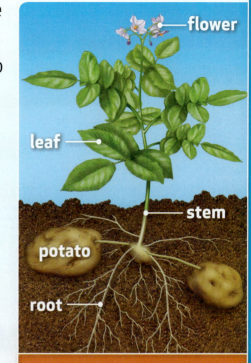

flower

leaf

stem

potato

root

Potatoes are stems that grow underground and store food for the plant.

How Plants Respond

Like all living things, plants respond to their environment. The **environment** of an organism is all the living and nonliving things around it. Temperature, light, and gravity are part of a plant's environment.

HOW A BEAN SEED GERMINATES
When the soil is warm enough, a bean seed will start to grow.

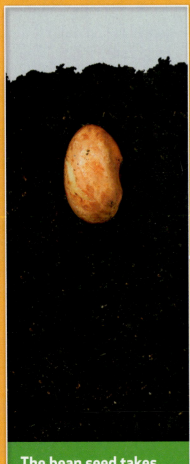

The bean seed takes in water from the soil and starts to swell.

The seed germinates, and the first root begins to grow downward.

The stem pushes upward. New roots grow.

Plants Respond to Heat Most plants grow only if the weather is warm enough. When the temperature rises in the spring, buds open and leaves begin to grow.

Many seeds respond to changes in the temperature of the soil. Some seeds germinate only when the soil is warm. When a seed **germinates** , it begins to grow. Because seeds do not germinate when it is cold, the young plants are not hurt by cold.

The seedling pushes out of the soil.
The first leaves start to unfold.

The first leaves open.

Plants Respond to Light Look at the plant in the big picture. Are its stems growing straight up? No, they are bending to the right. This plant is responding to the sunlight coming through the window. Plants get the energy they need by growing toward the light.

Some plants respond to changes in the direction of light during the day. Their leaves turn so they always face the sun. In the morning, they face east. In the evening, they turn to face the west. By facing the sun, their leaves get more energy from the sunlight.

These seedlings are growing toward the light coming from a lamp.

The stems of this plant are growing toward the sunlight coming through the window.

Plants Respond to Gravity Have you ever seen a plant with its leaves in the soil and its roots in the air? Of course not! Plants do not grow that way. Roots grow down and stems grow up.

Both roots and stems respond to gravity, but in opposite ways. Gravity is the force that pulls things toward Earth. Roots grow toward the pull of gravity. Stems grow away from gravity.

These corn stalks are growing away from the pull of gravity.

These responses to gravity help plants get the things they need. Growing down into the soil helps roots reach water and nutrients. Growing upward helps a stem reach sunlight.

When a seed germinates, gravity makes the young root grow down. Gravity also makes the young stem grow up and out of the soil.

Gravity causes the roots of this tree to grow down into the soil.

Before You Move On

1. How do stems respond to light?
2. Contrast the way roots respond to gravity with the way stems respond to gravity.
3. **Infer** In the early spring you plant some seeds. You water the seeds, but they do not grow. Why do you think the seeds did not germinate?

Groups of Plants

Flowering Plants Like all living things, plants reproduce. An organism **reproduces** when it makes more of its own kind.

Most plants reproduce with flowers. Look at the picture of the apple blossom below as you read about the parts of a flower. Flowers have male and female parts. The male parts of a flower make a fine powder called **pollen**.

FLOWER An apple blossom has male and female parts.

pollen

petal

male parts

female parts

seeds grow here

When bees or other animals visit flowers, pollen sticks to their bodies. When they visit other flowers, pollen rubs off on the female parts of the other flowers.

When pollen lands on the female part of a flower, a fruit begins to grow. Inside the fruit are one or more seeds. Each can become a young plant. The seed protects the young plant and contains a supply of food. The food helps the young plant start to grow.

POLLEN The male parts of a flower make pollen. Insects carry pollen to other flowers.

FRUIT AND SEEDS When pollen reaches the female part of a flower, an apple begins to grow. Inside the apple are several seeds.

seed

Plants with Cones How is a pine tree different from an apple tree? One important difference is that a pine tree has cones and an apple tree has apples! The seeds of a pine tree grow inside cones. Plants with cones do not have flowers or fruits.

Pine trees have two kinds of cones. Male cones make pollen. Wind blows the pollen to female cones. Then seeds grow inside the female cones. When the seeds are ripe, the scales of the female cones open and the seeds blow away.

These pitch pines are growing in Acadia National Park in Maine.

Like plants with flowers, plants with cones have roots, stems, and leaves. Many plants with cones have long thin leaves called needles. Pine needles stay on the tree all year, even in winter. Needles are covered by a thin layer of wax that protects them in cold weather.

male   female

Pitch pines have male and female cones. Male pine cones make pollen.

Seeds grow inside the female cones. When the cones open, the seeds blow away.

Plants Without Seeds Mosses and ferns do not have seeds. Instead, they reproduce with spores. A `spore` is a tiny part of a moss or fern that can grow into a new plant. Spores do not have a supply of food for the young plant.

Ferns usually have big leaves divided into many smaller parts. On the underside of the leaves are cases where spores grow.

The big, lacy leaves of ferns are called fronds.

Underneath the leaves of a fern are cases where spores grow.

Each case holds many tiny spores.

spores

Like ferns, mosses grow where it is shady and damp. Mosses do not have roots and stems, so they are usually much smaller than ferns. The spores of mosses form on tiny stalks that grow out of the green part of the plant.

The spores of mosses grow inside capsules at the top of slender stalks.

Before You Move On

1. What is pollen?
2. How are spores like seeds? How are they different?
3. **Infer** Fir trees have leaves shaped like thin needles. Do you think fir trees have flowers? Explain why.

Conclusion

Plants need food, water, nutrients, and space to live and grow. Leaves use water, air, and sunlight to make food. Roots take in water and nutrients. Stems support the plant and carry water and food between the roots and leaves. Plants respond to heat, light, and gravity. Many plants reproduce by making seeds in flowers or cones. Other plants reproduce by making spores.

Big Idea Plants have different parts that work together to help them live, grow, and reproduce.

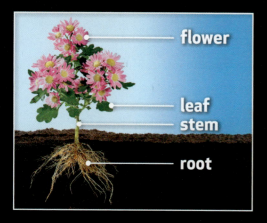

Vocabulary Review

Match the following terms with the correct definition.

A. organism	**1.** To begin to grow
B. environment	**2.** To make more of its own kind
C. germinate	**3.** A living thing
D. reproduce	**4.** A powder that is made by the male part of a flower or male cone
E. pollen	**5.** A tiny part of a moss or a fern that can grow into a new plant
F. spore	**6.** All the living and nonliving things around an organism

Big Idea Review

1. Describe Describe two ways that roots help a plant live and grow.

2. Identify The four main groups of plants are ferns, mosses, plants with cones, and plants with flowers. Which of these groups of plants reproduce with seeds? Which reproduce with spores?

3. Summarize How do the roots, stems, and leaves work together to make food for the plant? Begin by explaining what things a leaf needs to make food.

4. Classify A scientist discovers a new plant that has fruit with seeds inside. In what group of plants would the scientist classify this plant? Explain why.

5. Predict When the weather becomes warm in the spring, how will the buds on a tree respond?

6. Analyze How do stems and leaves respond to light? How does this response help a plant to grow?

Write About Plants

Infer Look at the picture. Why do you think the stem of the plant is bent? Write what you think happened to the plant. Also describe how the roots are growing. Explain why the stem and roots are growing this way.

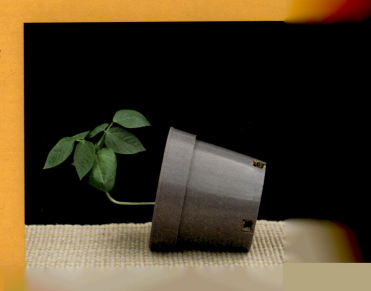

CHAPTER 1

LIFE SCIENCE EXPERT: BOTANIST

Grace Gobbo, Botanist

Do you enjoy learning about plants? Grace Gobbo does. She is a botanist, a scientist who studies plants. Her goal is to help save plants that are important to her community.

Grace lives in Tanzania in East Africa. Tanzania is a beautiful country with forests, grasslands, and mountains. It has more than 10,000 kinds of plants! Grace visits different places in Tanzania to learn about the plants that grow there.

TECHTREK
myNGconnect.com

Digital Library

Grace Gobbo studies trees, flowers, fruits, and vines. She is looking for plants that can cure diseases.

TECHTREK
myNGconnect.com

Student
eEdition

Digital
Library

Grace studies how people in her country use plants. She visits many different villages. There she asks the elders which plants they use to treat diseases. Then she writes down what the elders say. In the past, nothing was written down. Grace is helping to preserve their knowledge. Grace also works with people in the villages to help preserve the forests and other wild places.

These plants grow in tropical forests on Mount Kilimanjaro in Tanzania.

 Tanzania is a country in East Africa. Thousands of different plants grow there.

In the Forest of the Tallest Trees

Have you ever seen a forest with trees that are taller than 30-story buildings? You would if you visited the redwood forests of California and Oregon.

The Pacific Coast has just the right **environment** for redwood trees to grow. The ocean brings cool, rainy weather. Fog often fills the air.

Redwood trees grow in parts of California and Oregon.

environment

An **environment** is all the living and nonliving things around an organism.

TECHTREK
myNGconnect.com

e
Student
eEdition

Digital
Library

Redwoods are the tallest trees in the world. The trunk of a redwood has no elevators, but materials move up and down inside it. Its roots take in water and nutrients, which travel up the trunk to the leaves. The leaves use water, air, and sunlight to make food for the tree. Food travels from the leaves all the way down to the roots!

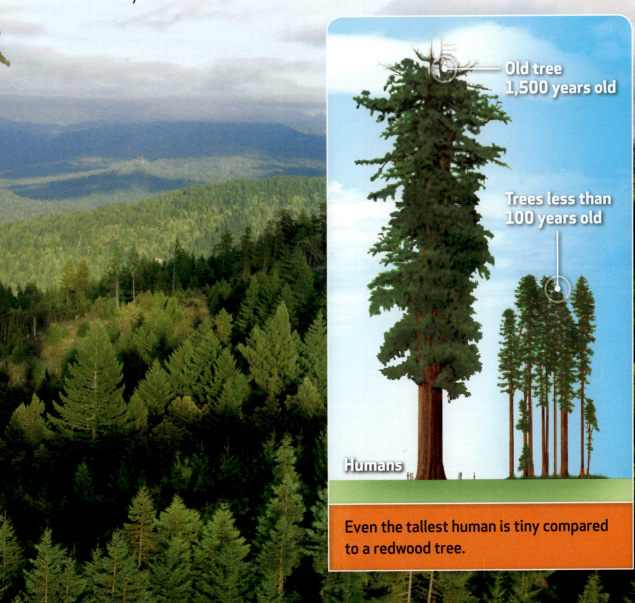

Old tree
1,500 years old

Trees less than
100 years old

Humans

Even the tallest human is tiny compared to a redwood tree.

Old Redwoods, New Redwoods

Many redwood trees are 500 years old. Some have lived for 2,000 years! Yet every redwood began life as a seed.

Redwood seeds grow inside cones. Redwood trees have two kinds of cones: male and female. Male cones make a powder called **pollen** . If pollen lands on a female cone, seeds will form.

Redwood trees have male and female cones. Seeds grow in the female cones.

female cones

male cones

TECHTREK
myNGconnect.com

Digital Library

If a seed lands in a good place, it will germinate.

pollen

Pollen is a powder made by a male cone or the male parts of a flower.

When the cone opens, its seeds fall to the ground. If a seed lands in a good place, it will start to grow, or **germinate** .

After many, many years, the seedling will grow into a tall redwood. The new tree will be similar to its parents. But each tree will also have its own special shape and pattern of branches.

After many years, a redwood tree will begin making cones.

germinate

When seeds **germinate,** they begin to grow into new plants.

Other Plants of the Forest

Many other kinds of plants also grow in the redwood forest. Most of these plants **reproduce** with flowers. The flowers of dogwoods and rhododendrons are very beautiful.

Rhododendrons flower in the spring.

reproduce

To **reproduce** is to make more of its own kind.

The seeds of flowering plants grow in fruits. The acorns of tanoaks are fruits. The sweet berries of huckleberry bushes are also fruits.

The shade of the tall trees makes the forest dark. But the damp forest floor is a good place for mosses and ferns to grow. Mosses and ferns reproduce by making **spores** .

Red huckleberries are a favorite food of bears and many birds.

Mosses and ferns often grow on fallen logs.

spore

A **spore** is a tiny part of a fern or moss that can grow into a new plant.

A Community High in the Trees

If you climbed up a redwood tree, what would you find? An amazing number of organisms. An **organism** is a living thing.

Stellar's jay

This trunk of a redwood is growing up from the branch.

leatherleaf fern

The marbled murrelet is a bird that feeds on fish in the ocean, but nests high up in old trees.

Salamanders look for food among the dead leaves.

organism

An **organism** is a living thing.

Soil has formed on some of the old branches. The art shows some of the plants and animals that live on these branches.

red huckleberry

yellow-cheeked chipmunk

A thick mat of soil has formed on this branch. The roots of many plants grow through it. Tiny animals called copepods live in the moist soil.

evergreen huckleberry

CHAPTER 1

SHARE AND COMPARE

Turn and Talk How do the roots, stems, and leaves of a redwood tree work together to help it live and grow? Work with a partner to form a complete answer to this question.

Read Select two pages from this section. Practice reading the pages. Then read them aloud to a partner. Talk about why the pages are interesting.

my SCIENCE notebook **Write** Write a conclusion about how the plants in the redwood forest live and grow. State what you think is the Big Idea of this section. Share what you wrote with a classmate. Compare your conclusions.

my SCIENCE notebook **Draw** Draw a picture of one plant or animal that lives in the redwood forest. Add labels or write a caption to explain what you drew. Combine your drawing with those of your classmates to make a mural of the redwood forest.

CHAPTER 2

HOW ARE ANIMALS ALIKE AND DIFFERENT?

You can find all sorts of animals in a pond, from slippery bullfrogs to delicate dragonflies. Some of the animals look similar. Others, such as the bullfrog and dragonfly, look very different. Scientists group animals by how they are alike and by how they are different. They study their body structures and how they have young.

One eye of this frog is larger than the head of the damselfly! How else are these two animals different? How are they alike?

TECHTREK
myNGconnect.com

In Chapter 2, you will learn:

FLORIDA NEXT GENERATION SUNSHINE STATE STANDARDS

SC.3.L.15.1 Classify animals into major groups (mammals, birds, reptiles, amphibians, fish, arthropods; vertebrates and invertebrates; those having live births and those which lay eggs) according to their physical characteristics and behaviors. **GROUPING ANIMALS; ARTHROPODS; FISHES, AMPHIBIANS, AND REPTILES; BIRDS AND MAMMALS; BABY ANIMALS**

SC.3.L.15.1 **Science in a Snap!** Classify animals into major groups (mammals, birds, reptiles, amphibians, fish, arthropods; vertebrates and invertebrates; those having live births and those which lay eggs) according to their physical characteristics and behaviors.

HOW ARE ANIMALS ALIKE AND

You can find all sorts of animals in a pond, from slippery bullfrogs to delicate dragonflies. Some of the animals look similar. Others, such as the bullfrog and dragonfly, look very different. Scientists group animals by how they are alike and by how they are different. They study their body structures and how they have young.

One eye of this frog is larger than the head of the damselfly! How else are these two animals different? How are they alike?

DIFFERENT?

SCIENCE VOCABULARY

classify (CLA-si-fī)

To **classify** is to place into groups based on characteristics. (p. 50)

Scientists classify crabs in the same group as spiders.

backbone (BAK-bōn)

A **backbone** is a string of separate bones that fit together to protect the main nerve cord in some animals. (p. 51)

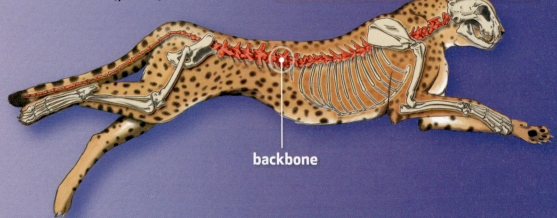

A cheetah's backbone bends when the cheetah runs.

backbone

my
Science Vocabulary

backbone
(BAK-bōn)

classify
(CLA-si-fī)

invertebrate
(in-VUR-tuh-brit)

vertebrate
(VUR-tuh-brit)

vertebrate (VUR-tuh-brit)

A **vertebrate** is an animal with a backbone. (p. 52)

A backbone helps many vertebrates walk, run, fly, jump, or swim.

invertebrate
(in-VUR-tuh-brit)

An **invertebrate** is an animal without a backbone. (p. 53)

A spiny lobster is an invertebrate.

Grouping Animals

Animals come in many shapes, sizes, and colors. How do scientists sort them? One way scientists **classify**, or group, animals is by their characteristics, or features and behaviors. These crabs all have ten legs and hard outer coverings. The iguanas have four legs and scaly skin. Classifying helps scientists better understand animals.

These iguanas and crabs live on the Galápagos Islands off the coast of South America. What characteristics do you observe about them?

One characteristic that scientists use to classify animals is whether or not an animal has a backbone. A **backbone** is a string of separate bones that fit together to protect the main nerve cord in some animals. The head and other bones are connected to the backbone. It helps the animal move.

These marine iguanas have a backbone. The Sally lightfoot crabs do not.

Animals with backbones are called **vertebrates**. The cheetah is just one kind of vertebrate. Catfish, tree frogs, king snakes, gulls, and bats are vertebrates, too. Even though each kind of animal is very different from the other, they all have backbones in about the same location as the cheetah's. The backbone is part of a vertebrate's skeleton.

TECHTREK
myNGconnect.com

Digital Library

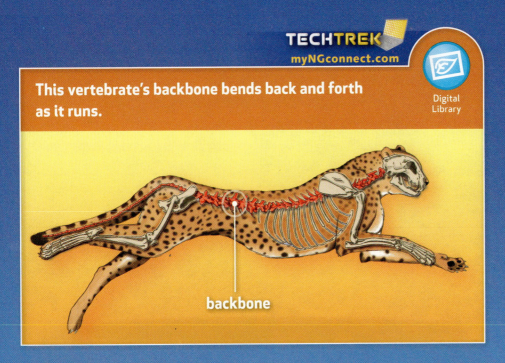

This vertebrate's backbone bends back and forth as it runs.

backbone

Animals without backbones are called **invertebrates**. Some invertebrates, such as earthworms, have no skeleton at all. Others such as clams grow a protective shell on the outside of their bodies. Still others such as lobsters and butterflies grow a hard outer covering that is like an outside skeleton. But none has a backbone.

An invertebrate's skeleton is on the outside of its body. It does not have a backbone.

Before You Move On

1. Why do scientists classify animals?
2. Observe the invertebrates and the vertebrates. How are they alike? How are they different?
3. **Apply** What are some kinds of animals that you think have a backbone?

Arthropods

You might think that vertebrates are the most common kinds of animals on Earth because you notice them more often. But more kinds of invertebrates called arthropods live on Earth than any other kind of animal! All arthropods have jointed legs, a body divided into sections, and a hard outside skeleton.

The fiddler crab has jointed legs like all other arthropods.

Scientists classify arthropods into three groups by the number of legs and body sections. Shrimp and crayfish belong to the same group as the fiddler crab. Ticks and scorpions belong to the same group as the golden silk spider. Honeybees and beetles belong to the same group as the ant.

This ant has a hard outside skeleton like all other arthropods.

This golden silk spider's body is divided into sections like all other arthropods.

Of all the different kinds of arthropods, most are insects. Scientists classify arthropods with three body sections and six legs attached to the middle section as insects. Most insects have wings, antennae for sensing the environment, and eyes made up of several smaller ones.

Find the spicebush swallowtail's antennae.

Notice that you can see through the wings on the bar winged skimmer.

This insect looks like a bumblebee, but it is really a bee beetle.

You might call all the insects on these pages "bugs." People commonly do. Some insects even have "bug" in their names. But they still may not be bugs. On these pages scientists would call only the milkweed bug a "bug." Scientists use the insect's wings and mouthparts to classify it as a bug.

The milkweed bug sucks juices from the milkweed plant's leaves and seeds.

Before You Move On

1. What is the largest group of arthropods?
2. Is an arthropod a vertebrate or invertebrate? Explain how you know.
3. **Analyze** How does the body structure of different arthropods help you classify them?

Fishes, Amphibians, and Reptiles

Fishes Scientists classify animals as fish using four characteristics. Fishes are vertebrates that live in water and have fins, scales, and gills. Fishes use gills to take in oxygen from the water. Fishes take in water through their mouths. Then the water passes out over the gills. As the water passes over the gills, oxygen moves from the water into the fish's blood.

TECHTREK
myNGconnect.com

Digital
Library

Sharks have gills and fins like all fishes.

gills

fin

Fishes can look very different from one another. A shark is a fish. Sharks have scales that are rough as sandpaper. You can see the slits where the water moves out over the gills. Most fishes are like the koi. Its scales feel slimy. It also has a covering over the gills so you can't see the gills on the outside of the fish's body.

Koi have scales and live in water like all fishes.

Amphibians   Have you ever seen a frog or a toad? Frogs and toads are part of another group of vertebrates called amphibians. Newts and salamanders are also part of this group. Amphibians spend part of their lives in water. They also live part of their lives on land. When they are fully grown, most amphibians have four legs and thin, damp skin.

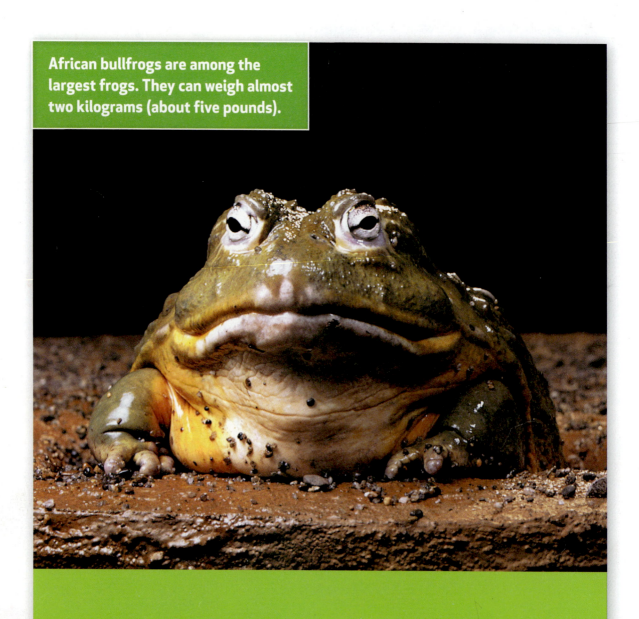

African bullfrogs are among the largest frogs. They can weigh almost two kilograms (about five pounds).

Most amphibians hatch in the water from eggs. Young amphibians live in water and breathe with gills. Their bodies change as they grow into adults. They grow legs, and their tails may shrink. Most amphibians also grow lungs. With these changes, amphibians can live on land.

Eastern, or red-spotted, newts lay eggs in water.

The young newts have feathery gills and live in water.

This newt is almost an adult. It has grown legs and lungs, and lives on land.

Reptiles You may have seen members of another group of vertebrates called reptiles. Snakes, turtles, and lizards are all reptiles. Alligators and crocodiles are reptiles, too.

Unlike fish and amphibians, reptiles mostly lay their eggs on land. Reptile eggs have a soft but tough shell that protects them. When the eggs hatch, the young reptiles look like their parents. They have lungs and breathe air.

Compare this collared lizard to the reptiles on the next page. How are they alike? How are they different?

Reptiles have a covering of hard scales that is not slimy. Scales help keep their bodies from drying out. The scales also protect reptiles from other animals that may try to eat them. Scales can also help reptiles move.

Like other snakes, the emerald tree boa has special scales on its belly that help it climb trees.

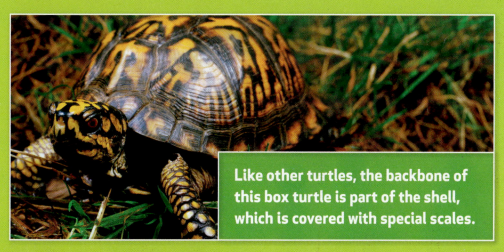

Like other turtles, the backbone of this box turtle is part of the shell, which is covered with special scales.

Before You Move On

1. What are the characteristics of a fish?
2. What characteristic do fish and young amphibians have in common?
3. **Generalize** How are a newt and a lizard alike? How are they different?

Birds and Mammals

Birds What do you notice about the vertebrates on this page? While they look very different, they both share certain characteristics. Scientists classify vertebrates with feathers, two legs covered with scales, and wings as birds.

The beak and legs of the scarlet ibis are longer than many birds. It wades in water to catch food.

No other kind of animal has feathers. These help birds fly and stay warm. Both birds puff out their feathers to trap air when they are cold. The trapped air next to their skin warms and helps the birds stay warm. But not all birds fly. The feathers of a flying bird such as the ibis have a different structure than the feathers of a penguin, which doesn't fly.

Even though all birds have wings, they do not all fly. The rockhopper penguin hops to where it needs to be!

Mammals Vertebrates that have hair or fur and make milk to feed their young are called mammals. You cannot always see the hair or fur on a mammal. Some mammals, such as whales, lose their hair soon after they are born.

Most mammals take care of their young after they are born. Female mammals make milk for their young until the young are old enough to eat other foods.

Elephants are mammals so they have hair. But the hair is not as important for keeping warm as in other mammals.

Mammals, too, are very different from one another. Some have long trunks, some live in the ocean, and some can fly! Scientists use many characteristics and behaviors to classify mammals.

Science in a Snap! **Classify Animals**

Enrichment Activities

Observe the pictures above. Classify each animal as a vertebrate or an invertebrate. Organize your classifications in a chart.

Share your work with your partner.

Before You Move On

1. What characteristic do birds have that no other vertebrate has?
2. What two characteristics make mammals different from other vertebrates?
3. **Apply** Name five other kinds of animals that you think are mammals.

Baby Animals

Scientists observe how animals reproduce. Most kinds of animals have young by laying eggs. The young animals develop inside the eggs. Soon after hatching, most of these young are able to survive on their own. Arthropods can lay hundreds of eggs at one time. Most fishes, amphibians, reptiles, and birds lay eggs, too.

The mother emu laid the eggs. Then the father emu kept them warm until they hatched about eight weeks later.

Most mammals do not lay eggs. The young develop inside the mother and are born alive. But the young cannot take care of themselves. They drink milk from their mother until they can eat other foods. Many times the young learn behaviors from the parents, such as how to find food and how to survive on their own.

For the next three weeks the mother will hide the fawn in tall grass. She will return twice a day to feed it milk.

Before You Move On

1. Where can baby animals develop?
2. How is milk important to mammals?
3. **Draw Conclusions** Dolphins are a large animal that lives in the ocean. They have young that are born alive. Do you think dolphins are most likely a fish or a mammal? Explain.

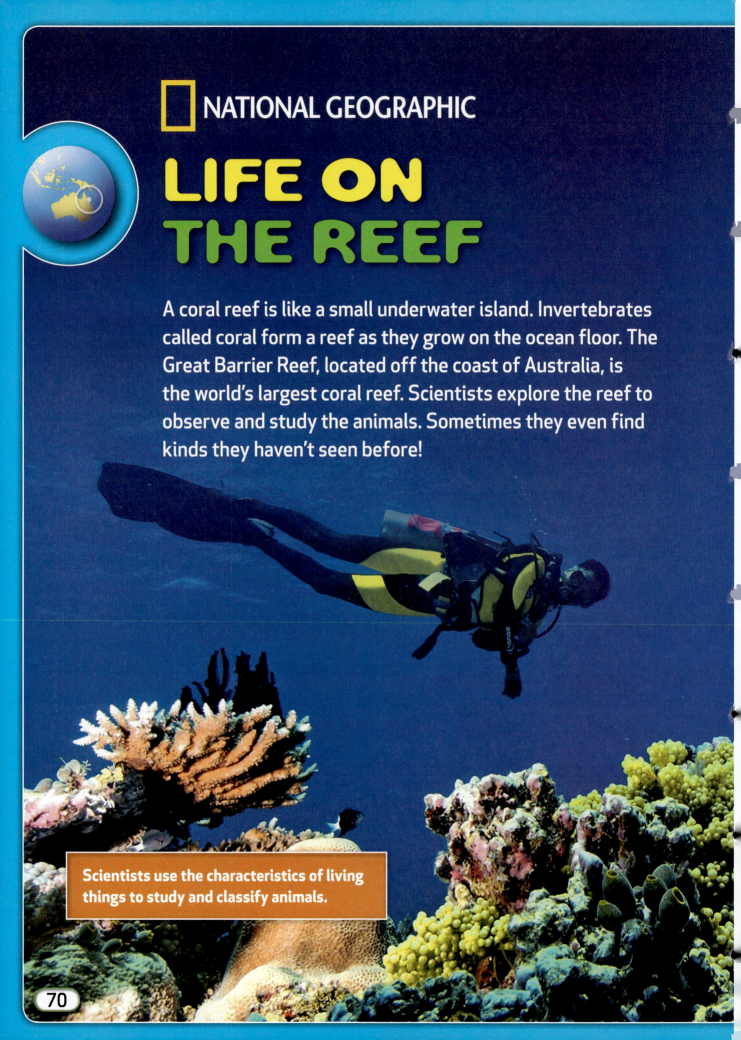

NATIONAL GEOGRAPHIC

LIFE ON
THE REEF

A coral reef is like a small underwater island. Invertebrates called coral form a reef as they grow on the ocean floor. The Great Barrier Reef, located off the coast of Australia, is the world's largest coral reef. Scientists explore the reef to observe and study the animals. Sometimes they even find kinds they haven't seen before!

Scientists use the characteristics of living things to study and classify animals.

Many different kinds of animals live on a coral reef. Scientists use cameras and other tools to observe the animals. Animals can be hard to find because they hide in the coral and live in holes left when coral dies. Scientists try to classify any new kinds of animals they haven't seen before by comparing their characteristics with groups of living things they already know.

This reptile is a sea snake. What are some characteristics of it?

What arthropod characteristics do you see in this green-banded snapping shrimp?

What animals might scientists compare this sea horse with?

Conclusion

Classifying animals helps scientists to study and better understand them. They observe animals' characteristics. Invertebrates are animals that do not have backbones and vertebrates are animals with backbones. Arthropods are the largest group of invertebrates. Vertebrates include fishes, amphibians, reptiles, birds, and mammals. Some animals reproduce by laying eggs while others have young that are born alive.

Big Idea Animals can be classified into groups based on their characteristics and behaviors.

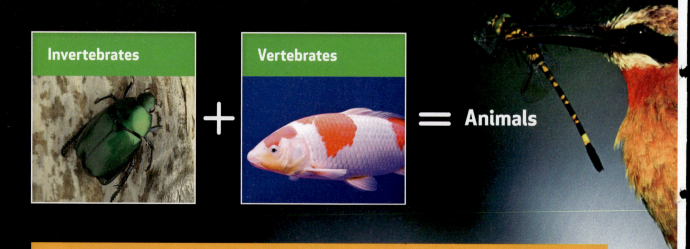

Vocabulary Review

Match the following terms with the correct definition.

A. classify   **1.** An animal without a backbone

B. vertebrate   **2.** A string of separate bones that fit together to protect the main nerve cord in some animals

C. invertebrate

D. backbone   **3.** An animal with a backbone

                    4. To place into groups based on characteristics

Big Idea Review

1. **Name** What are some characteristics of arthropods?

2. **Define** What is a vertebrate?

3. **Explain** Why is it useful to classify animals?

4. **Compare and Contrast** Compare how young develop in arthropods, fishes, amphibians, reptiles, birds, and mammals.

5. **Draw Conclusions** Bats are animals that fly. They are covered in hair and make milk for young. Do you think they are birds or mammals? Give reasons.

6. **Generalize** The word *amphibian* comes from old words that mean "double" and "life." Why is this a good name for frogs, toads, salamanders, and newts?

Write About Animals

my SCIENCE notebook

Classify This young deer is probably still drinking milk from its mother. Write a few sentences that tell what characteristics you could use to classify both animals. Then tell what groups you would classify them in.

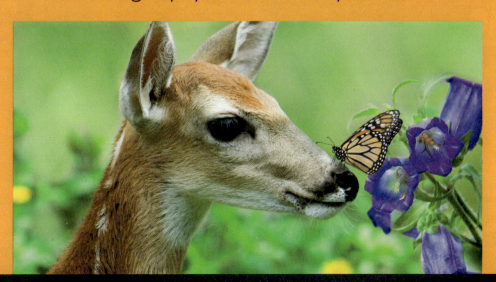

LIFE SCIENCE EXPERT: VETERINARIAN

Carlos Sanchez, Zoo Veterinarian

A veterinarian—or vet, for short— is a doctor for animals. A vet helps animals stay healthy and treats them when they are sick. In school, vets study all about animals. They learn about the diseases that animals get and how to treat those diseases. Many vets do surgery too.

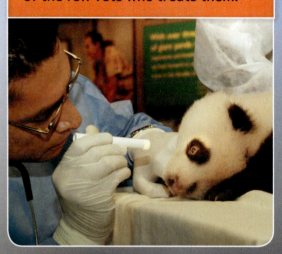

Pandas live in only a few zoos in the United States. Dr. Sanchez is one of the few vets who treats them.

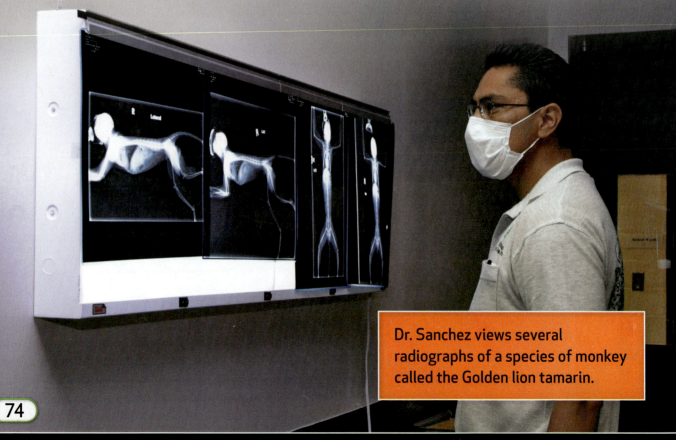

Dr. Sanchez views several radiographs of a species of monkey called the Golden lion tamarin.

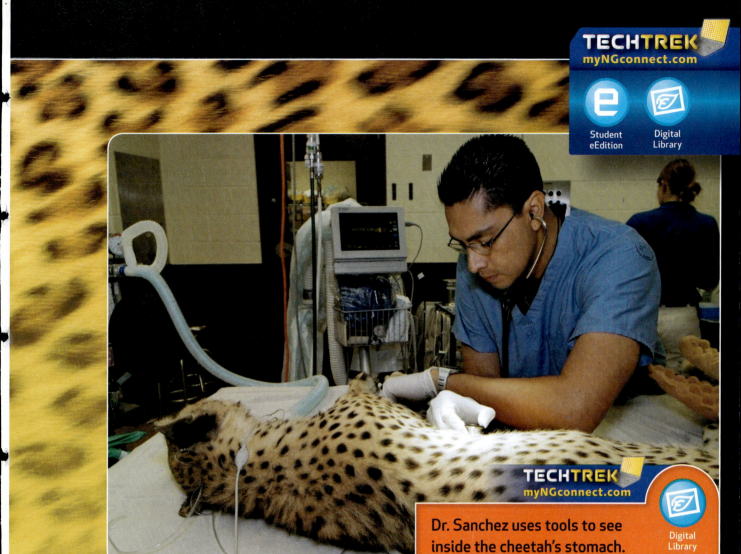

Dr. Sanchez uses tools to see inside the cheetah's stomach.

Many vets treat cats, dogs, and other pets. Some vets care for farm animals, such as horses and cows. But Dr. Carlos Sanchez treats more unusual animals. He is a vet at the Smithsonian's National Zoo in Washington, D.C. He treats all of the zoo's animals.

Dr. Sanchez does not work only at the zoo. He has traveled to China to study pandas. He has traveled to Africa to study giraffes and cheetahs. Dr. Sanchez shares what he learns so that everyone benefits from his trips.

When Dr. Sanchez was in grade school, he knew he wanted to be a vet someday. His advice for young people is to study hard. Maybe they will operate on an elephant some day!

The Florida Everglades:
More Than Meets the Eye

The Everglades is a national park that covers the southern tip of Florida. The land here is flat, wet, and has many different habitats, or areas where animals can live. This environment supports many kinds of animals.

Everglades National Park is often called the "river of grass."

Everglades National Park

Invertebrates called arthropods live in the Everglades. Insects are one type of arthropod. Like all arthropods, insects have a body divided in sections, jointed legs, and a hard outer skeleton. Insects have 3 body sections and 6 legs. They can be further classified into groups based on other traits such as mouth parts.

SOME EVERGLADES **ARTHROPODS**

Zebra longwing butterfly

Black salt march mosquito

Lubber grasshopper

invertebrate

An **invertebrate** is an animal without a backbone.

Birds

All of the animals of the Everglades have characteristics that give clues as to how scientists can **classify** them. The purple gallinule, or swamp hen, is covered in brightly colored feathers. Its large feet help it walk on floating plants.

The roseate spoonbill has pink feathers and two long legs. It looks for small fish, frogs, or other animals. When it finds one, it eats it with a snap of its spoon-shaped beak! Feathers, two legs, and beaks are characteristics that all birds have.

Why do you think this bird is called a spoonbill?

The purple gallinule also swims on the surface of water like a duck.

classify

To **classify** is to place into groups based on characteristics.

Reptiles

Alligators and crocodiles are the largest reptiles in the Everglades. Like birds, reptiles are **vertebrates**, or animals with **backbones**.

The Everglades is the only place in nature where alligators and crocodiles live side by side. How can scientists tell them apart? One way is to observe the snout, or front of the head. Both have nostrils on top of the snout so the animals can breathe while they float in the water. But observe how the shapes of the snouts are different.

TECHTREK
myNGconnect.com

Digital Library

The snouts of American crocodiles are more pointed.

American alligators have more rounded snouts.

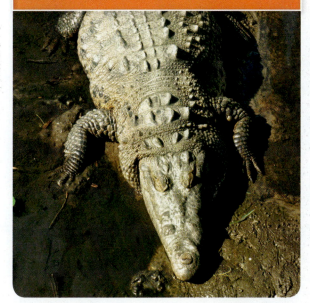

vertebrate

A **vertebrate** is an animal with a backbone.

backbone

A **backbone** is a string of separate bones that fit together to protect the main nerve cord in some animals.

Fishes

The alligator gar is not an alligator at all, but a fish! Its thick scales, covers over its gills, and long snout must have made it look like an alligator to the people who started calling it that. The alligator gar is one of the biggest freshwater fish in North America. It can grow to be 3 meters (10 feet) and over 45 kilograms (100 pounds)!

The sawfish has more characteristics like a shark than a gar. Sawfish live in saltier water where the Everglades meets the ocean.

Alligator gars live further inland where the water has little salt in it.

Amphibians

The swamps and ponds of the Everglades make a great home for amphibians, such as the pine woods tree frog. Frogs, like almost all amphibians, live part of their life in water and part of their life on land. The tadpole stage of the pine woods tree frog lives in water. The adult frog spends time on land. This frog's sticky toes help it climb on palm fronds and tree limbs in this watery environment.

This young pine woods tree frog lives in water and has gills. It will grow lungs and legs later in its life.

The adult pine woods tree frog has legs and lungs like other amphibians. It also has sticky toe pads that help identify it as one of a group called tree frogs.

sticky toe pad —

You can see that the American green tree frog also has sticky toe pads.

Mammals

Several mammals live in the Everglades. They all have hair or fur and make milk for their young that are born alive. But they can be very different from one another. Some eat plants, some eat insects, and still others eat other vertebrates. Only a few of some kinds are left, such as the Florida panther. It has characteristics much like panthers or cougars that live elsewhere. But the Florida panther lives only in southern Florida.

American mink

Southern flying squirrel

WHERE ARE THE FLORIDA PANTHERS?

Look at the brown area in the maps. A long time ago, large numbers of Florida panthers used to roam Florida and other parts of the southeastern United States. Today, scientists think only about 100 panthers are alive. They live in a small area of southern Florida, and their young are born only in the Everglades.

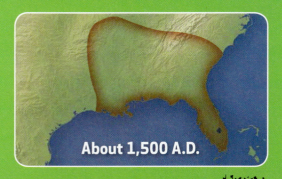

About 1,500 A.D.

Today

Young take about three months to develop inside the mother panther. She will have one to three kittens at a time and lets them drink her milk for about a year.

CHAPTER 2

SHARE AND COMPARE

Turn and Talk How can the different kinds of animals in the Everglades be classified? Form a complete answer to this question together with a partner.

Read Select two pages in this section. Practice reading the pages so you can read them smoothly. Then read them aloud to a partner. Talk about why the pages are interesting.

my SCIENCE notebook **Write** Write a conclusion that tells the important ideas about what you have learned about animals in the Everglades. State what you think is the Big Idea of this section. Share what you wrote with a classmate. Compare your conclusions.

my SCIENCE notebook **Draw** Form groups of six. Have each person draw a different kind of animal that lives in the Everglades. Label the group you are drawing an example of. Label the characteristics you used to classify the animal. Combine the drawings to show the different kinds of animals in the Everglades.

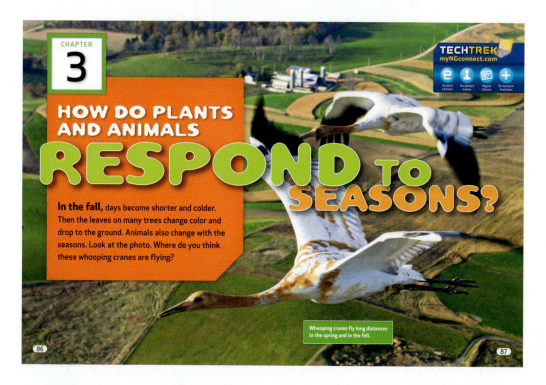

CHAPTER

3

HOW DO PLANTS
AND ANIMALS

RESPOND TO
SEASONS?

In the fall, days become shorter and colder.
Then the leaves on many trees change color and
drop to the ground. Animals also change with the
seasons. Look at the photo. Where do you think
these whooping cranes are flying?

TECHTREK
myNGconnect.com

Student   Vocabulary   Digital   Enrichment
eEdition   Games   Library   Activities

Whooping cranes fly long distances
in the spring and in the fall.

86   87

In Chapter 3, you will learn:

FLORIDA NEXT GENERATION SUNSHINE STATE STANDARDS

SC.3.L.17.1 Science in a Snap! Describe how animals and plants respond to changing seasons.

PLANTS CHANGE DURING THE SEASONS, ANIMALS CHANGE DURING THE SEASONS,
OTHER SEASONAL CHANGES

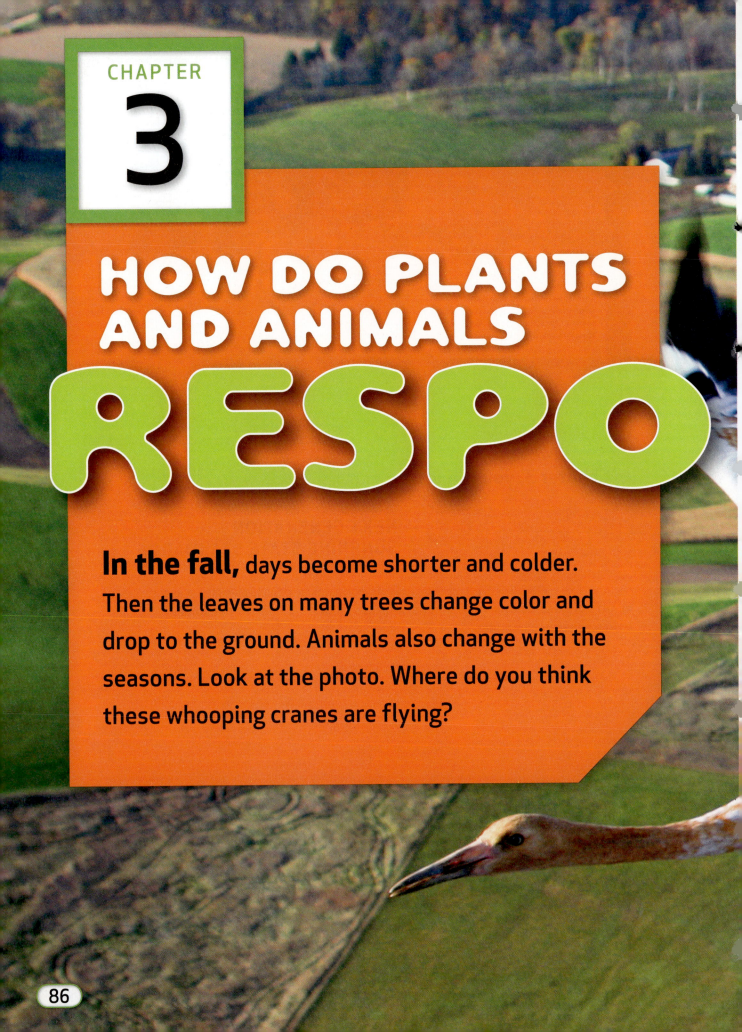

HOW DO PLANTS AND ANIMALS RESPO

In the fall, days become shorter and colder. Then the leaves on many trees change color and drop to the ground. Animals also change with the seasons. Look at the photo. Where do you think these whooping cranes are flying?

ND TO SEASONS?

Whooping cranes fly long distances in the spring and in the fall.

SCIENCE VOCABULARY

season (SĒ-zun)

A **season** is a time of year with certain weather patterns and day lengths. (p. 90)

Winter is the coldest season.

deciduous (di-CIJ-yu-us)

A **deciduous** plant sheds its leaves every year. (p. 90)

This deciduous tree has no leaves in the winter.

evergreen (EV-ur-grēn)

An **evergreen** is a plant that keeps its green leaves all year. (p. 91)

These evergreen trees keep their thin leaves, or needles, through the winter.

my Science Vocabulary

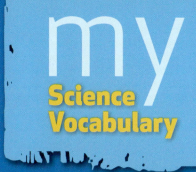

deciduous
(di-CIJ-yu-us)

evergreen
(EV-ur-grēn)

hibernate
(HĪ-bur-nāt)

migrate
(MĪ-grāt)

season
(SĒ-zun)

TECHTREK
myNGconnect.com

Vocabulary
Games

hibernate (HĪ-bur-nāt)

When animals **hibernate**, they go into a state that is like a deep sleep during cold winter months. (p. 97)

Chipmunks hibernate through the winter.

migrate (MĪ-grāt)

When animals **migrate**, they move to another place to meet their basic needs. (p. 98)

Monarch butterflies migrate in the fall.

Plants Change During the Seasons

Changes in the Way Plants Look Have you ever noticed how trees change during the <mark>seasons</mark>? A season is a time of year with certain weather patterns and day lengths.

Look at the winter scene below. The trees without leaves are <mark>deciduous</mark>. Deciduous trees shed their leaves in the fall. First, sugars and other nutrients in the leaves are moved to the rest of the tree. Then the leaves change color and fall to the ground. New leaves will grow in the spring.

> Deciduous trees lose their leaves in fall.

> Evergreen trees keep their green leaves all winter.

WINTER
Maple leaves would freeze in winter. Hard scales protect the buds.

SPRING
As the weather warms, new maple leaves grow from buds.

Evergreen trees do not lose their leaves every year. Many evergreen trees have long thin leaves called needles. A layer of hard wax protects the needles in winter.

Science in a **Snap!**

Comparing Deciduous and Evergreen Leaves

aspen leaf

spruce leaves

Describe the shape of the aspen leaf and the spruce leaves. How are the aspen and spruce leaves similar? How are they different? Which of these leaves do you think is evergreen? Why?

SUMMER
Maple leaves stay green all summer long. They make food for the tree.

FALL
Maple leaves change color and dry out. Soon they fall from the tree.

Changes in the Way Plants Grow Many plants, such as cornflowers, live for only one year. In the spring, their seeds germinate, or begin to grow. During the summer, they flower and seeds begin to form. The plants die in the fall, but leave many seeds behind. Their seeds lie in the soil over winter. The seeds will begin growing in the spring.

cornflower

Some plants live for many years. Trees have woody stems that help them survive cold weather. Other plants have thin stems that die back in cold weather, but their roots stay alive. Lilies and tulips have bulbs. Bulbs grow underground. Bulbs store food that the plants use to begin growing in the spring.

A LILY THROUGH THE SEASONS

WINTER The plant survives cold weather as a bulb underground.

FALL Food moves down into the bulb and the leaves die.

SUMMER The plant flowers and forms seeds.

SPRING The bulb sends up a stem and leaves.

Before You Move On

1. What is an evergreen plant?
2. How do deciduous trees survive cold winter weather?
3. **Infer** A marigold dies after it makes seeds. Why might marigolds keep appearing in the same field?

Animals Change During the Seasons

Changes in the Way Animals Look To live through cold weather, many animals change in ways that help them keep warm. Deer, rabbits, and other animals grow heavy fur coats or thick layers of body fat. Some birds grow more feathers. When temperatures rise in the spring, the animals shed the extra fur or feathers. This helps them stay cool.

Musk oxen grow thick, heavy undercoats for winter.

In summer, musk oxen shed their undercoats.

Some animals change color with the seasons. In the summer, their fur or feathers are dark. This color blends in with the rocks, soil, and plants around them. It is hard for other animals to see them. In the winter, these animals turn white. Their white coats blend in with the snow.

TECHTREK
myNGconnect.com

Digital Library

In summer, arctic foxes have brown coats. In winter, they grow white coats. How does this change in color help the foxes live?

Changes in the Way Animals Behave It is difficult for many animals to find food in winter when leaves are gone and the ground is covered with snow. Some animals spend most of the fall gathering food and storing it. They eat that food during the winter.

Other animals eat extra food in the fall. This food is stored in their bodies as fat. Their bodies use the fat for energy during the cold winter. The extra layer of fat is also like a thick coat that keeps them warm.

This chipmunk is gathering food and will store it in its underground nest.

During cold winter months, some animals hibernate. When animals hibernate, they go into a state that is like a deep sleep. Their bodies use less energy when they hibernate. While they are asleep, the animals live on the fat stored in their bodies.

Many animals, such as chipmunks, find shelter and sleep much of the winter. From time to time they wake up to eat the food they stored in their nests.

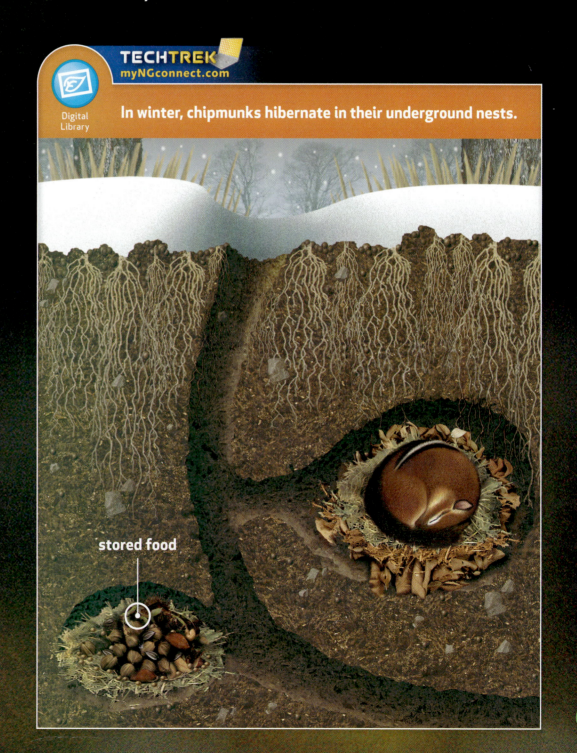

TECHTREK
myNGconnect.com

Digital Library

In winter, chipmunks hibernate in their underground nests.

stored food

Some kinds of animals **migrate** when the seasons change. To migrate is to move to another place to meet basic needs. Many birds spend the summer in Canada and the northern United States. There they raise their young. In the fall the birds migrate to warmer places in the south. In the spring they return north. Migrating helps birds get the food they need.

Scarlet tanagers spend the winter in South America. In spring they return to the forests of eastern North America to raise their young.

Many insects live through the winter as eggs or larvae. Others burrow deep in the soil. But monarch butterflies migrate south in the fall. Moving to a warmer place helps the monarchs survive.

TECHTREK
myNGconnect.com

Enrichment Activities

Each fall monarch butterflies migrate south to California, Mexico, or Florida. In the spring monarchs fly north.

CANADA

UNITED STATES

MEXICO

Monarchs need to eat during their long flight south. This monarch is sipping nectar from a wildflower.

Before You Move On

1. List two ways animals may look different in summer and winter.
2. How does migrating help animals survive?
3. **Analyze** Some years oak trees do not make many acorns. Then there is less food for the animals. How might this affect an animal that hibernates?

Other Seasonal Changes

In many places, summers are warm or hot, and winters are cold. But weather does not change like this everywhere. For example, the African savanna has warm weather all year long. There the weather changes in other ways. Parts of the year are rainy, and other parts are dry.

In eastern Africa, great herds of wildebeests migrate in search of grass and water.

The baobab tree has leaves during the rainy season.

In the dry season, the baobab loses its leaves. This helps the tree save water.

During the rainy season on the African savanna, grasses and other plants grow well. Herds of wildebeests and zebras eat the grass. During the dry season, grasses do not grow. Then the herds migrate to other places where there is grass to eat and water to drink.

Before You Move On

1. How does the savanna change during the rainy season?
2. How are the seasonal changes of a baobab tree like those of a maple tree? How are they different?
3. **Predict** What would happen to the wildebeests if they did not migrate?

Manatee Migration IN FLORIDA

Manatees are sometimes called sea cows. Like cows, manatees are large animals that eat plants. They spend much of their time looking for plants to eat in shallow rivers, marshes, and ocean channels. Hydrilla and water hyacinths are some of the manatees' favorite foods.

Manatees can move easily between the fresh water of rivers and the salt water of the ocean. However, they cannot live in cold water. This is why many manatees migrate.

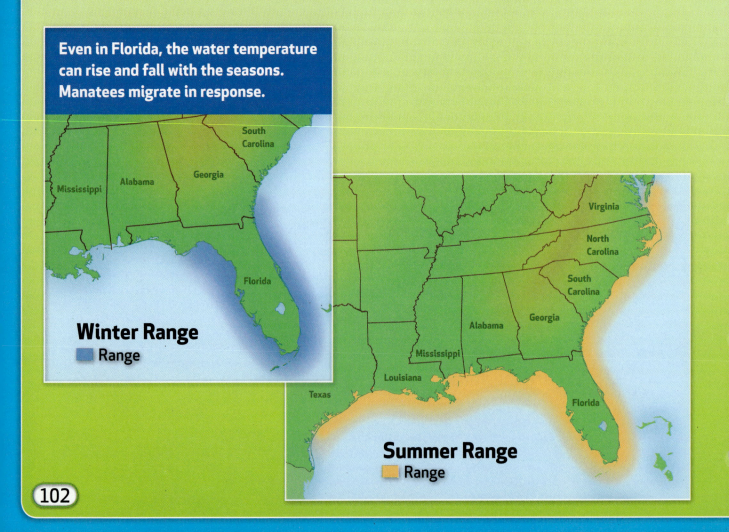

Even in Florida, the water temperature can rise and fall with the seasons. Manatees migrate in response.

Winter Range
■ Range

Summer Range
■ Range

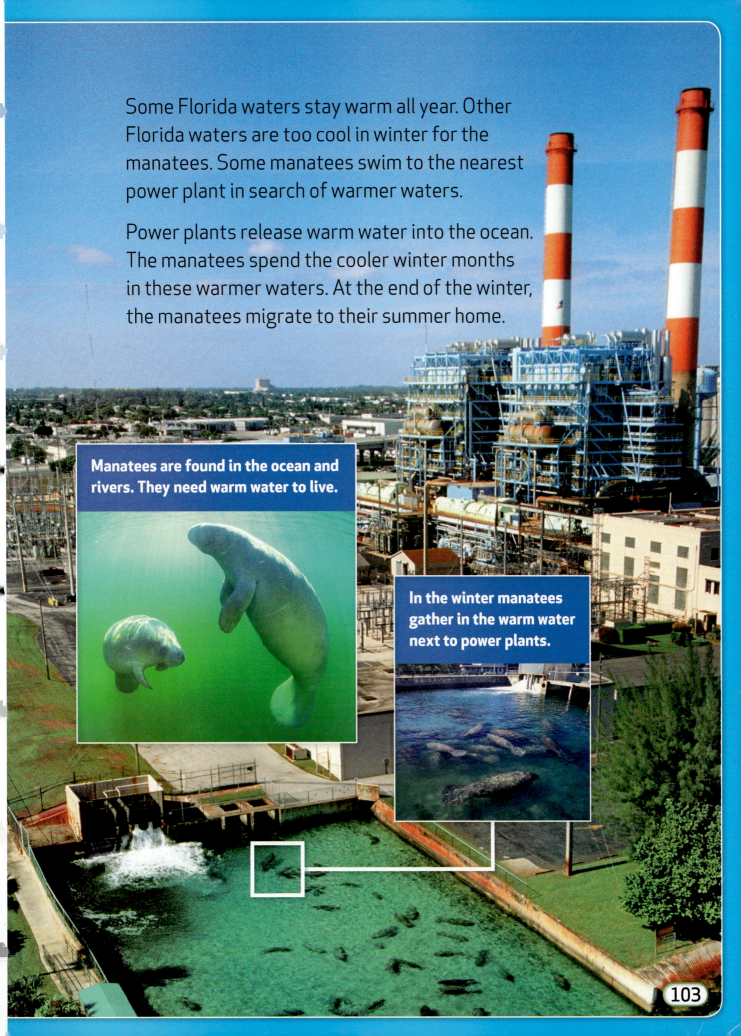

Some Florida waters stay warm all year. Other Florida waters are too cool in winter for the manatees. Some manatees swim to the nearest power plant in search of warmer waters.

Power plants release warm water into the ocean. The manatees spend the cooler winter months in these warmer waters. At the end of the winter, the manatees migrate to their summer home.

Manatees are found in the ocean and rivers. They need warm water to live.

In the winter manatees gather in the warm water next to power plants.

Plants and animals respond to changing seasons. Deciduous plants shed their leaves during cold or dry seasons. Many animals grow and shed fur or feathers as the seasons change. Some animals change color. Some animals store food for the winter. Others eat extra food and build up fat. Some animals hibernate in winter. Others migrate as seasons change.

Big Idea Plants and animals have many different ways to live through changing seasons.

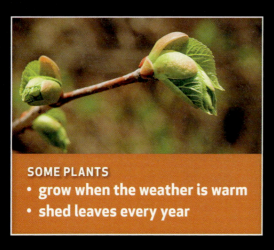

SOME PLANTS
- **grow when the weather is warm**
- **shed leaves every year**

SOME ANIMALS
- **change how they look**
- **change how they behave**

Vocabulary Review

Match the following terms with the correct definition.

A. evergreen

B. season

C. deciduous

D. hibernate

E. migrate

1. To go into a state that is like a deep sleep during cold winter months

2. A plant that keeps its green leaves all year

3. To move to another place to meet basic needs

4. A tree that sheds its leaves every year

5. A time of year with certain weather patterns and day lengths

Big Idea Review

1. Describe Describe two ways that an animal's body may change to help it stay warm in winter.

2. Identify What are the main seasons in the African savanna? In which season is there more food for wildebeests?

3. Explain How does changing color help an arctic fox live through the winter?

4. Sequence List these stages of a deciduous plant in order: leaves grow, branches are bare, leaves change color, buds open, leaves fall. Begin with winter.

5. Generalize Why is it more difficult for most animals to live in the winter than in the summer?

6. Predict People sometimes build roads or fences across animals' migration paths. How might these structures affect migrating animals?

my SCIENCE notebook

Write About Seasons

Explain Daffodils grow and bloom in the spring. How does making a bulb help daffodils live through a cold winter?

CHAPTER 3 LIFE SCIENCE EXPERT: ULTRALIGHT PILOT

JOE DUFF

Operation Migration

Joe Duff helped found Operation Migration. Operation Migration uses ultralight aircraft to teach captive whooping cranes to migrate from Wisconsin to Florida.

What is your job with Operation Migration?

I am in charge of the team that trains the birds to follow our ultralight aircraft. As a pilot, I get to lead the birds on their first migration south.

What is the best part of your job?

Flying with the birds. They migrate in the fall when the colors are bright and the air is cool. We fly over hills and towns and down valleys. I can look out and see a long line of birds gliding off the wingtip. Then I realize how beautiful they are.

Have there been any surprises in your job?

No one could have predicted that we could use tiny airplanes to teach birds where to migrate. We were surprised when the first birds began to follow our aircraft. We were surprised when they came back on their own the next spring.

TECHTREK
myNGconnect.com

Student
eEdition

Digital
Library

What's been your greatest accomplishment?

In 2006, a whooping crane chick was hatched at the Necedah National Wildlife Refuge in Wisconsin. It was the first wild chick to hatch in the Midwest in over 100 years. That chick followed its parents to Florida along the route we showed them. Since then, it has returned every year.

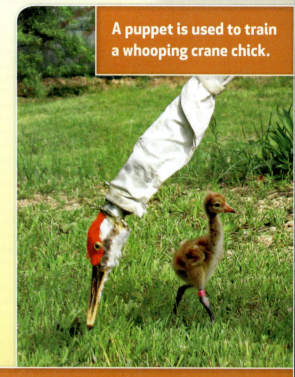

A puppet is used to train a whooping crane chick.

TECHTREK
myNGconnect.com

Digital
Library

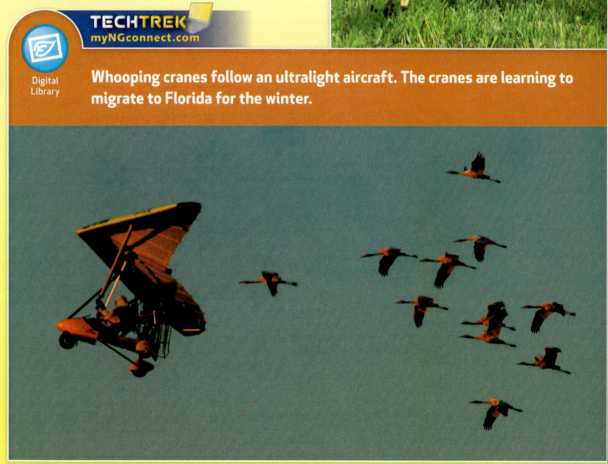

Whooping cranes follow an ultralight aircraft. The cranes are learning to migrate to Florida for the winter.

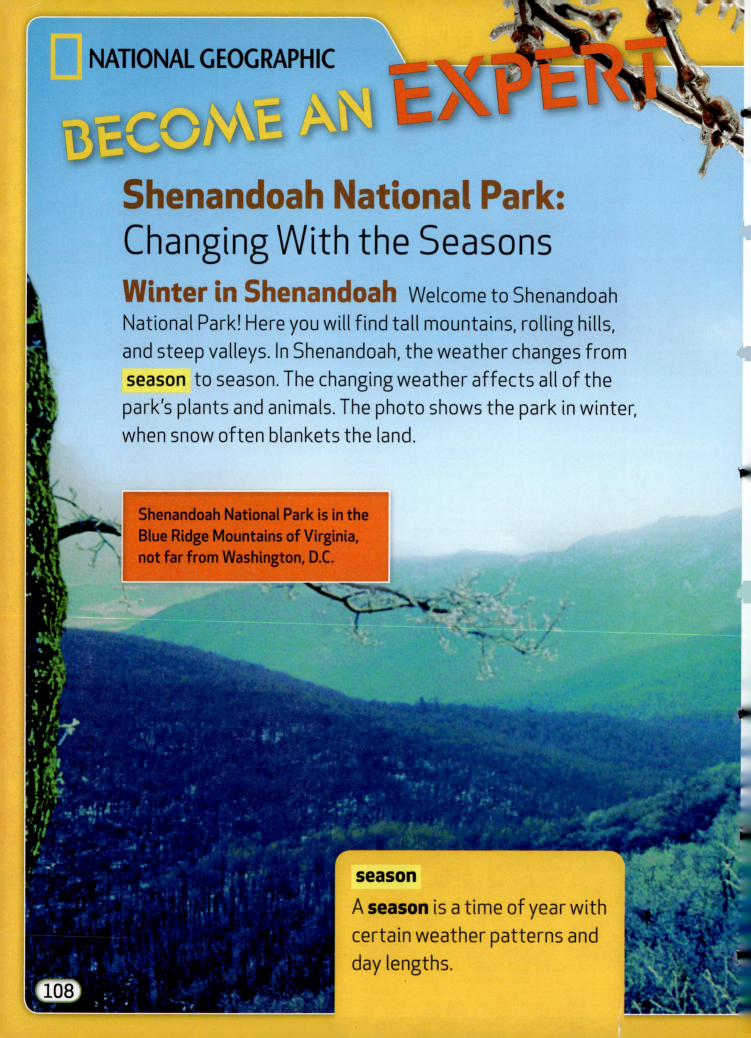

BECOME AN EXPERT

Shenandoah National Park: Changing With the Seasons

Winter in Shenandoah Welcome to Shenandoah National Park! Here you will find tall mountains, rolling hills, and steep valleys. In Shenandoah, the weather changes from **season** to season. The changing weather affects all of the park's plants and animals. The photo shows the park in winter, when snow often blankets the land.

Shenandoah National Park is in the Blue Ridge Mountains of Virginia, not far from Washington, D.C.

season

A **season** is a time of year with certain weather patterns and day lengths.

TECHTREK
myNGconnect.com

Student
eEdition

Digital
Library

Some animals, such as woodpeckers and squirrels, are active all winter. Bobcats hunt for rabbits, squirrels, and mice. The bobcats' thick fur keeps them warm.

In winter you would not see chipmunks or woodchucks. These animals **hibernate**, or go into a state that is like a deep sleep.

Bobcat

Pileated woodpecker

hibernate

When animals **hibernate,** they go into a state that is like a deep sleep during cold winter months.

Spring in Shenandoah

Spring brings warm weather to Shenandoah. Snow melts and rain falls. In the woods you can see wildflowers, such as violets, spring beauties, trilliums, and lady's slippers. You'll also see the new leaves of **deciduous** trees, such as oaks and hickories.

TECHTREK
myNGconnect.com

Digital Library

Many wildflowers, such as these yellow lady's slippers, bloom in the spring.

deciduous

A **deciduous** tree sheds its leaves every year.

Deer and many other animals of the Shenandoah have offspring, or young, during the spring. The young animals will find plenty of food to help them grow during the warm months ahead. Spring also brings the return of many birds. These birds **migrate** to warmer places during the winter. In the spring they come back to nest and raise their young.

The cerulean warbler spends the winter in South America. It returns to the Shenandoah in April or May.

new leaves

migrate

When animals **migrate,** they move to another place to meet their basic needs.

Summer in Shenandoah

Summer is the warmest season in the park. Plants grow quickly in the warm weather. The plants provide food to many animals. Deer nibble on leaves and stems. Bees, butterflies, and many other insects visit flowers, getting nectar and pollen.

SPOTTED SKUNK

BOX TURTLE

Some animals, such as bats, owls, and skunks, come out at night. Skunks have a keen sense of smell that helps them hunt. They defend themselves with a foul-smelling spray.

Different plants and animals live in different parts of the park. In the Big Meadows region there are many ponds and marshes. During the summer, frogs hop in and out of the water, insects buzz overhead, and raccoons visit to catch fish.

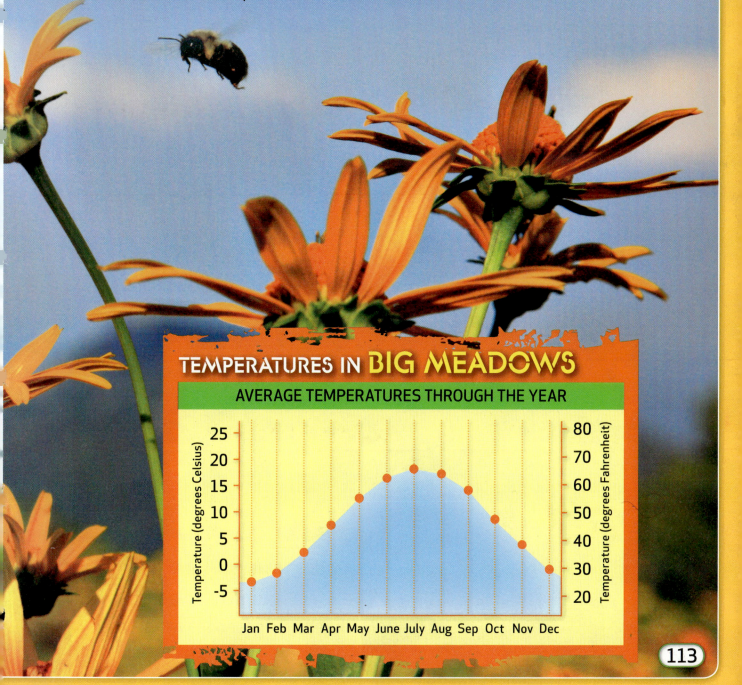

TEMPERATURES IN BIG MEADOWS

AVERAGE TEMPERATURES THROUGH THE YEAR

Temperature (degrees Celsius) — left axis: 25, 20, 15, 10, 5, 0, -5

Temperature (degrees Fahrenheit) — right axis: 80, 70, 60, 50, 40, 30, 20

Months: Jan Feb Mar Apr May June July Aug Sep Oct Nov Dec

Fall in Shenandoah

When fall arrives, the weather becomes cool. The fruits of many plants are now ripe. Wild grapes hang from the vines. Many bushes have red or black berries.

The leaves of deciduous trees change color, and then fall to the ground. But the leaves of **evergreen** trees, such as pines, do not fall off. They will stay on the trees through the winter.

evergreen

An **evergreen** is a plant that keeps its green leaves all year.

The animals of the park are busy getting ready for winter. Squirrels gather hickory nuts and acorns. Bears eat fruits and acorns, as well as insects and fish. They store this food in their bodies as fat. Soon the bears will find dens where they will sleep through the winter. The fat in their bodies will keep the bears warm while they sleep.

When fall arrives, black bears store fat for winter. Then they find dens where they will sleep.

CHAPTER 3

SHARE AND COMPARE

Turn and Talk   How do the changing seasons in Shenandoah affect the animals that live there? Work with a partner to form a complete answer to this question.

Read   Select two pages from this section. Practice reading the pages. Then read them aloud to a partner. Talk about why the pages are interesting.

my SCIENCE notebook   **Write**   Write a conclusion that tells the important ideas you learned about the seasons in Shenandoah. State what you think is the Big Idea of this section. Share what you wrote with a classmate. Compare your conclusions.

my SCIENCE notebook   **Draw**   Choose a season. Draw a plant or animal from the park as it may appear during this season. Add labels or write a caption to help explain what you drew. Share your drawing with your classmates. Group your drawings according to season.

FLORIDA

EARTH SCIENCE

What Is Earth Science?

Earth science investigates all aspects of our home planet from its changing surface, to its rocks, minerals, water, and other resources. It also includes the study of Earth's atmosphere, weather and climates. As Earth is an object in space, Earth science also includes the study of Earth's relationship with the sun, moon, and stars. People who study our planet are called earth scientists.

You will learn about these aspects of earth science in this unit:

WHAT PROPERTIES CAN YOU OBSERVE ABOUT THE SUN?

The sun is a star—a ball of hot, glowing gases. Its energy moves throughout the solar system, causing objects on Earth to heat up. Its gravity works with that of Earth to keep Earth in orbit. Earth scientists study the sun and its effects on Earth.

WHAT CAN YOU OBSERVE ABOUT STARS?

The sky is filled with tiny points of light. But each one can be much larger than the sun. Stars have different properties—brightness, size, color, and temperature. These properties, along with how far away they are, can make stars look different from each other. Earth scientists study the stars using tools such as telescopes.

MEET A SCIENTIST

Madhulika Guhathakurta:
Astrophysicist

Madhulika Guhathakurta is an astrophysicist and the lead program scientist for NASA's Living With a Star (LWS) program. LWS focuses on understanding and ultimately predicting how the sun changes and what the effects of those changes are here on Earth. Simply put, Lika's research is about understanding "space weather" better.

Lika is also leading a worldwide effort known as the International Living With a Star (ILWS) program. This initiative is made up of all the space agencies of the world to contribute towards the scientific goal for space weather understanding.

"Our sun is just one of billions of stars in the universe. However, the sun is important because it's the closest star to Earth. Without the sun, life on Earth would not exist. Without the sun, Earth would be frozen."

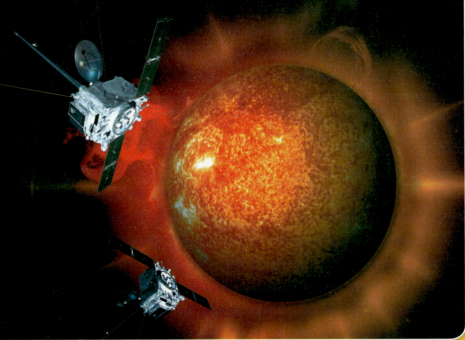

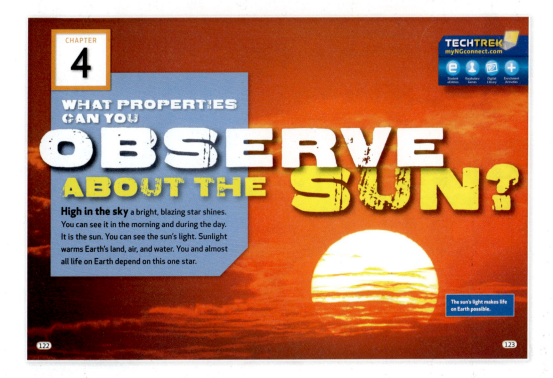

CHAPTER
4

WHAT PROPERTIES
CAN YOU

OBSERVE
ABOUT THE SUN?

High in the sky a bright, blazing star shines. You can see it in the morning and during the day. It is the sun. You can see the sun's light. Sunlight warms Earth's land, air, and water. You and almost all life on Earth depend on this one star.

TECHTREK
myNGconnect.com

The sun's light makes life on Earth possible.

122   123

In Chapter 4, you will learn:

FLORIDA NEXT GENERATION SUNSHINE STATE STANDARDS

SC.3.E.5.2 Identify the sun as a star that emits energy; some of it in the form of light. **PROPERTIES OF THE SUN, THE SUN IS A SOURCE OF ENERGY**

SC.3.E.5.3 Recognize that the sun appears large and bright because it is the closest star to Earth. **PROPERTIES OF THE SUN**

SC.3.E.5.4 Explore the Law of Gravity by demonstrating that gravity is a force that can be overcome. **GRAVITY**

SC.3.E.6.1 Demonstrate that radiant energy from the sun can heat objects and when the sun is not present, heat may be lost. **THE SUN IS A SOURCE OF ENERGY**

SC.3.E.6.1 Science in a Snap! Demonstrate that radiant energy from the sun can heat objects and when the sun is not present, heat may be lost.

WHAT PROPERTIES CAN YOU OBSE ABOUT THE

High in the sky a bright, blazing star shines. You can see it in the morning and during the day. It is the sun. You can see the sun's light. Sunlight warms Earth's land, air, and water. You and almost all life on Earth depend on this one star.

RVE SUN?

The sun's light makes life on Earth possible.

SCIENCE VOCABULARY

sun (SUN)

The **sun** is the star that is nearest to Earth. (p. 126)

The sun looks bigger and brighter than other stars because it is the nearest star to Earth.

energy (EN-ur-jē)

Energy is the ability to do work or cause a change. (p. 128)

The sun sends energy out into space.

light (LĪT)

Light is a kind of energy you can see. (p. 129)

The Earth gets light energy from the sun.

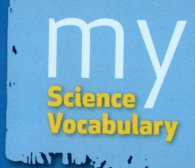
my Science Vocabulary

energy
(EN-ur-jē)

sun
(SUN)

gravity
(grav-u-tē)

temperature
(TEM-pur-ah-chur)

light
(LĪT)

transform
(trans-FORM)

TECHTREK
myNGconnect.com

Vocabulary Games

transform (trans-FORM)

To **transform** is to change. (p. 131)

Light energy can transform into heat energy and warm the land and other objects.

temperature (TEM-pur-ah-chur)

Temperature is a measure of how hot or cold something is. (p. 133)

The sidewalk has a cool temperature in the shade.

gravity (GRAV-u-tē)

Gravity is a force that pulls objects toward each other. (p. 136)

Earth's gravity pulls a thrown ball back to Earth.

125

Properties of the Sun

The sun is just one of millions of stars in the sky. Why does it look bigger and brighter than other stars? The sun looks so large and bright because it is the nearest star to Earth. It is about 150 million kilometers (93 million miles) away. If you could drive to the sun in a car, it would take you about 177 years!

The Mars Rover took this photograph of the sun from the surface of Mars. The sun appears much smaller from Mars than it does from Earth. That's because Mars is farther away from the sun.

The sun is a medium-sized star. Some stars are much larger. Others are smaller. Compared to Earth, however, the sun is huge. More than 1 million Earths could fit inside it! As you can see in the model below, the sun is much bigger than even the largest planet, Jupiter.

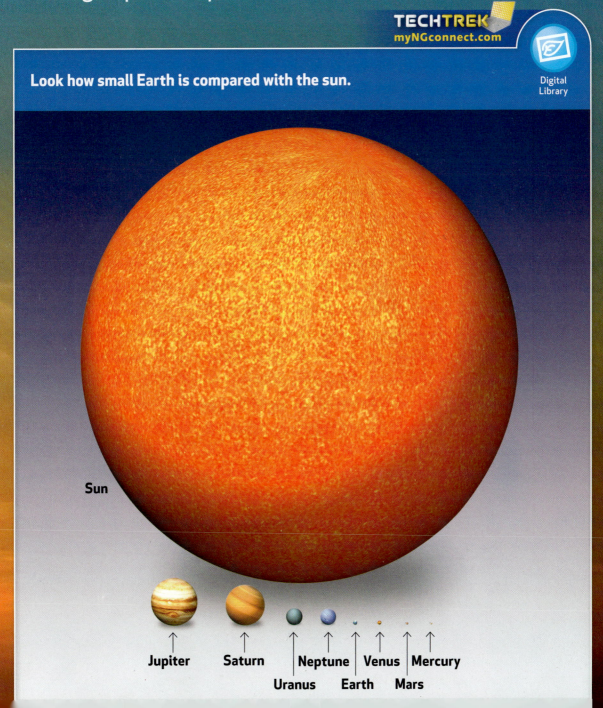

Look how small Earth is compared with the sun.

Sun

Jupiter Saturn Neptune Venus Mercury

Uranus Earth Mars

Life on Earth depends on **energy** from the sun. Energy is the ability to do work or cause a change. You need energy to live. The sun's energy is everywhere and changes form. The sun's energy is in the fuel used to run cars. It's in the food you eat. It's even in the air that moves around you as wind.

The sun is very bright. Do not look at it directly, or it will damage your eyes.

Different kinds of energy come from the sun. Some of the energy is in the form of light . Light is a kind of energy you can see. During the day, you can see sunlight all around you. When sunlight reaches Earth it warms things. You can feel the sunlight warming your skin when you walk outside.

Before You Move On

1. Name a kind of energy that comes from the sun.
2. Why does the sun look bigger and brighter than other stars?
3. **Draw Conclusions** Why do you need the sun to live?

The Sun as a Source of Energy

Energy from the sun goes out into space. It moves out in all directions as light. Sunlight reaches throughout the solar system. The sun provides Earth and all the other planets with energy.

The sun provides energy to all the planets in the solar system, but not in equal amounts. Look at the diagram. Which planet do you think receives more of the sun's energy, Earth or Neptune? Why do you think so?

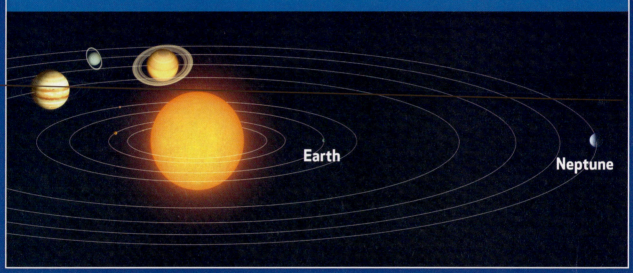

Earth

Neptune

By providing light energy to Earth, the sun also provides heat. When light energy from the sun strikes Earth, it **transforms**, or changes, to heat energy. This heat warms the ground. It also then warms the air above the ground.

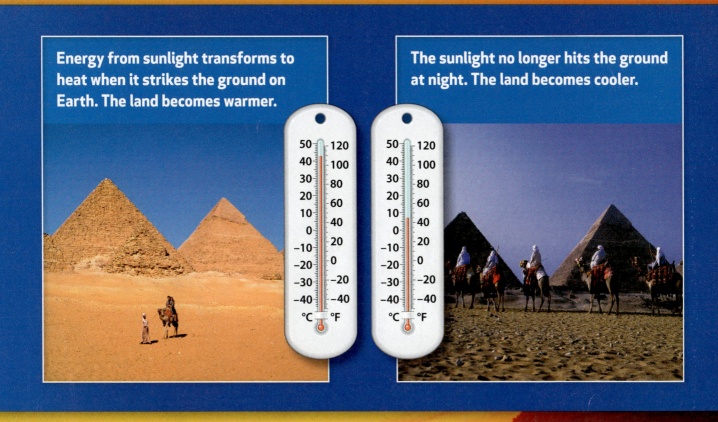

Energy from sunlight transforms to heat when it strikes the ground on Earth. The land becomes warmer.

The sunlight no longer hits the ground at night. The land becomes cooler.

Have you walked on a blacktop surface on a sunny summer afternoon? It's hot! Light energy from the sun has transformed to heat energy. It has warmed the blacktop.

Imagine walking on the same blacktop in the late evening. It's much cooler. The sun no longer shines on it. The blacktop has lost heat energy.

The long shadows show that it is late afternoon in this picture. Do you think that the blacktop road is hotter or cooler than it was at midday?

Trees provide shady spots on the black path. Is the surface hotter in the sun or in the shade?

The **temperature** of the blacktop changes during the day. Temperature is a measure of how hot or cold something is. When sunlight hits an object, heat energy causes that object's temperature to go up.

Science in a Snap! Compare Temperatures

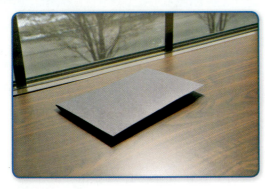

Fold two sheets of black paper in half. Place one in the sun and one in the shade.

Wait 20 minutes. Place a thermometer inside each sheet of paper. Wait 1 minute. Record the temperatures.

What can you infer about how sunlight changes an object?

Before You Move On

1. What is temperature?
2. What effect does sunlight have on an object?
3. **Predict** The sun is shining on a piece of blacktop at 10 a.m. It is still shining on the same piece at 4 p.m. Will the blacktop be hotter in the morning or the afternoon? Why?

SUNSPOTS
STORMS ON THE SUN

From Earth, you see the outer layer of the sun. This layer is so bright that you cannot see past it to the inner layers. But not all of the sun's surface is bright. Dark blotches swirl and move across it. These sunspots are storms on the sun. Sunspots are darker than the rest of the surface because they are much cooler.

Even the smallest sunspots are more than 1,600 kilometers (1,000 miles) across. The largest are bigger than Earth.

Sunspots come and go. They grow and shrink. They appear alone or in large groups. Sunspots follow a pattern. The sun has a 22-year cycle, or pattern, of sunspots. During this cycle the number of sunspots increases and then decreases.

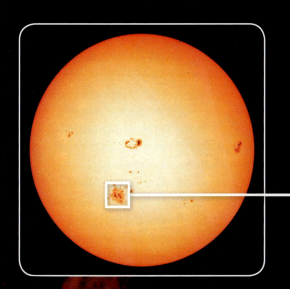

This picture shows a close-up of a sunspot.

The sunspot cycle begins with just a few sunspots. Over the next 11 years, their number grows. At the peak of the cycle, the sun may have as many as 100 sunspots. Through the next 11 years, many disappear. Then the cycle begins again.

Gravity

Gravity is a force that pulls objects toward each other. Gravity between the sun and Earth pulls Earth toward the sun. Then why doesn't Earth fall into the sun? Think about twirling a ball connected to a string. You start the ball moving by throwing it out in a straight line. Then right away you pull on the string and the ball. That pull keeps the ball moving in a circle around you.

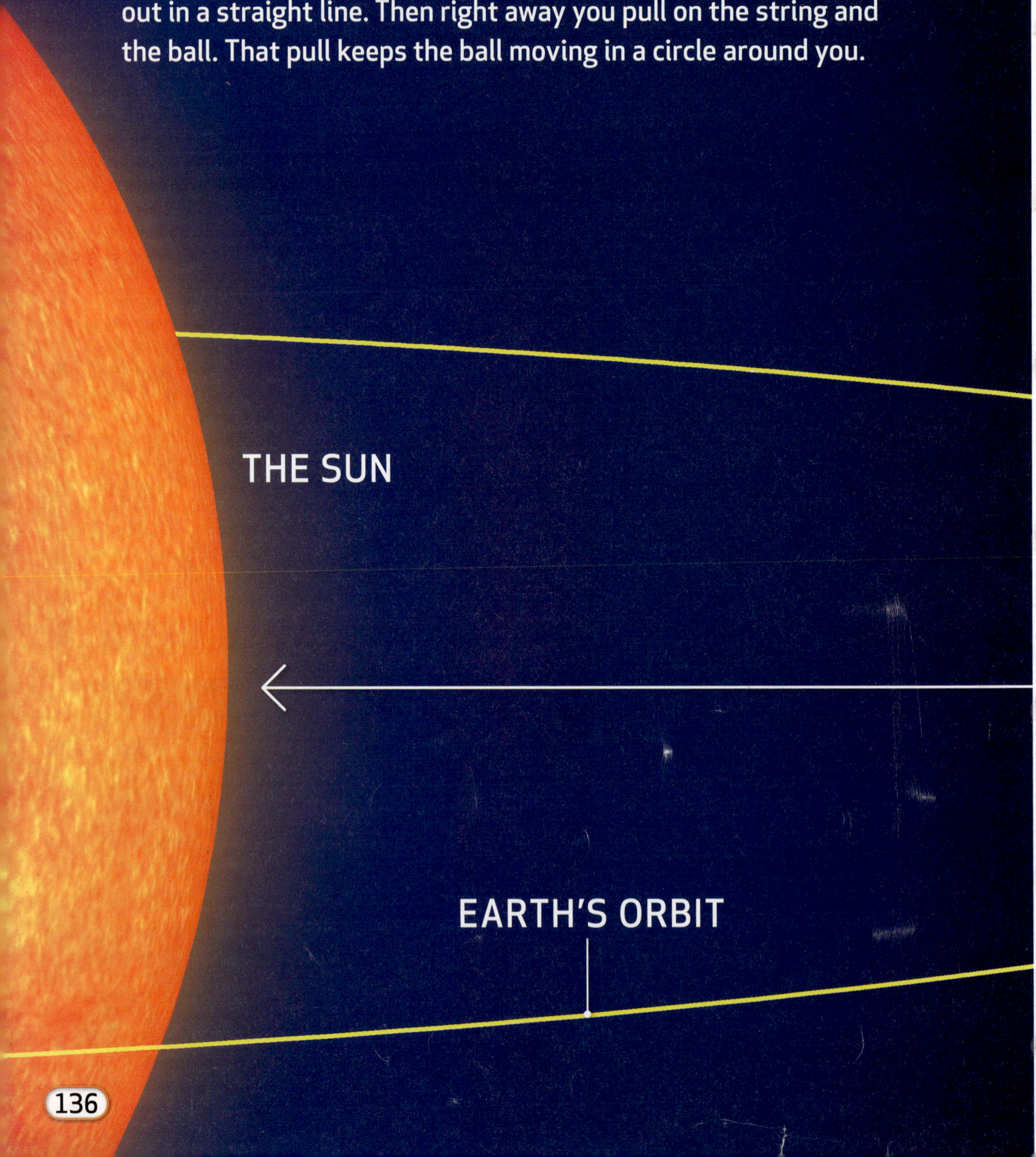

THE SUN

EARTH'S ORBIT

The same thing happens with the sun and Earth, but with no strings. Without gravity, Earth would fly through space in a straight line. The sun's gravity pulls Earth. Earth's gravity pulls the sun. The force of gravity keeps Earth in orbit around the sun.

EARTH

If the sun's gravity stopped pulling on Earth, Earth would fly straight out of its orbit.

When one object has more mass than another, its gravity is stronger. Look at the ball in the picture. Earth has a lot more mass than the ball. So, Earth's gravity is much greater than the ball's gravity. When you throw a ball up, you use a force to overcome Earth's gravity. But the force of Earth's gravity is still greater, and down . . . down comes the ball toward Earth.

GREATER MASS = STRONGER FORCE OF GRAVITY

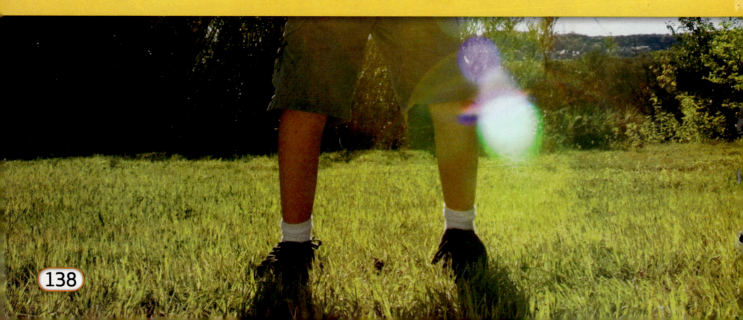

Then how can a rocket blast into space without falling back to Earth? Like a ball, a rocket has force to help it overcome Earth's gravity. Unlike a ball, the rocket gets far away from Earth fast. When an object is farther away, the pull of Earth's gravity is weaker. The rocket gets to a point where Earth's gravity is too weak to pull it back.

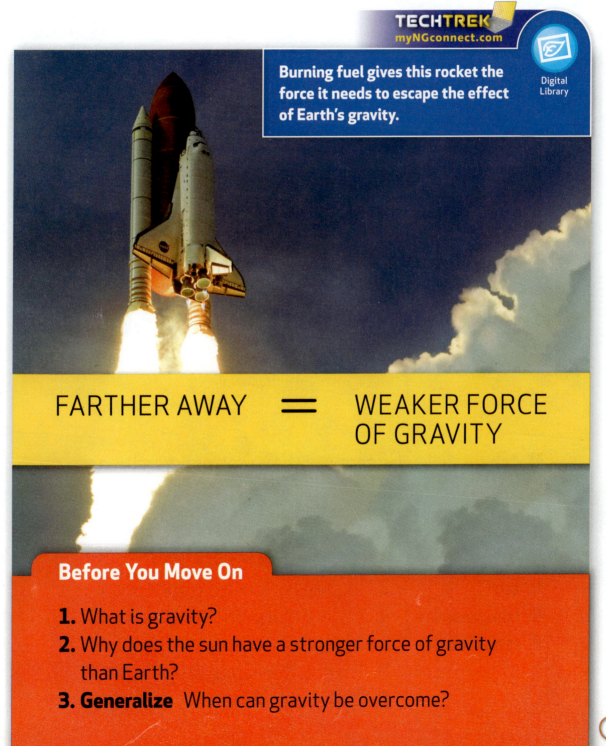

Burning fuel gives this rocket the force it needs to escape the effect of Earth's gravity.

FARTHER AWAY   =   WEAKER FORCE OF GRAVITY

Before You Move On

1. What is gravity?
2. Why does the sun have a stronger force of gravity than Earth?
3. **Generalize** When can gravity be overcome?

The sun is the closest star to Earth. So the sun looks brighter and larger than other stars. The sun produces energy. Some of this energy is light that can transform to heat. Gravity keeps Earth in orbit around the sun. The sun's gravity is stronger than Earth's because it has more mass.

Big Idea The sun produces energy that provides light and heat to Earth. Its gravity holds Earth in orbit.

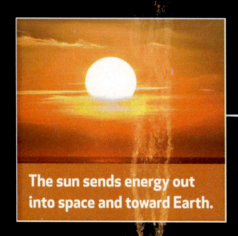

The sun sends energy out into space and toward Earth.

Life on Earth uses light and heat energy from the sun.

Vocabulary Review

Match the following terms with the correct definition.

A. energy

B. gravity

C. light

D. sun

E. temperature

F. transform

1. A kind of energy you can see
2. A measure of how hot or cold something is
3. To change
4. A force that pulls objects toward each other
5. The star that is nearest to Earth
6. The ability to do work or cause a change

Big Idea Review

1. Define What kind of energy does Earth get from the sun?

2. Recall The sun is just one star you see in the sky. Why does it appear so much bigger than the other stars?

3. Cause and Effect Suppose one bike is lying in the sun. Another bike is lying in the shade of a tree. Which bike will be cooler? Why?

4. Explain What happens when sunlight hits Earth?

5. Generalize Explain how the temperature of an object changes at different times during the day.

6. Infer Think of a way you can overcome the force of gravity. Explain what you had to do.

Write About the Sun

Explain Millions of years from now, the sun will produce less energy. Write about what you think will happen to Earth.

CHAPTER
4

EARTH SCIENCE EXPERT: SOLAR ENGINEER

How can people use the sun's energy? Ask a solar engineer.

Much of the sun's energy bounces back into space. But what if people could collect more of the sun's energy? Then they could meet their energy needs without harming the environment. Solar engineer Chuck Kutscher is working to make this happen.

TECHTREK
myNGconnect.com

Digital Library

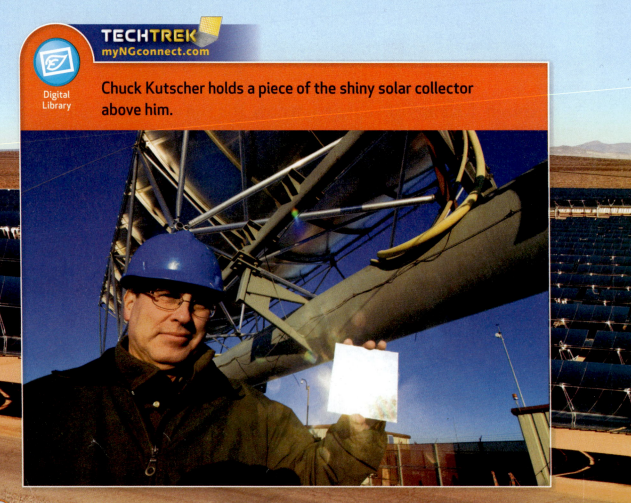

Chuck Kutscher holds a piece of the shiny solar collector above him.

TECHTREK
myNGconnect.com

Student
eEdition

Digital
Library

Kutscher leads a team of engineers. Together they use computers to help design new solar collectors. The solar collectors catch the sun's energy. They focus its energy on tubes, which become heated. People can use this heat energy to make electricity.

Kutscher loves solar energy research. Why? Solar energy, or energy from the sun, is friendly to the environment. It does not dirty the air, as burning coal and oil do. But that is not the only thing Kutscher likes about his work. Talking about solar energy to people is "the fun part of my job," he says.

To become a solar engineer, Kutscher studied math, science, and engineering. A student should also "be interested in exploring and learning new things," he says. "Wanting to help solve the world's energy problems is important, too."

These solar energy collectors have shiny, curved surfaces. They reflect sunlight onto a tube.

NATIONAL GEOGRAPHIC

BECOME AN EXPERT

Solar Light Shows:
The Sun's Energy Colors the Sky

Earth depends on the **sun** for many things. The sun's **gravity** helps keep Earth in orbit around the sun. **Energy** from the sun enables life to exist. But the sun can also provide colorful entertainment.

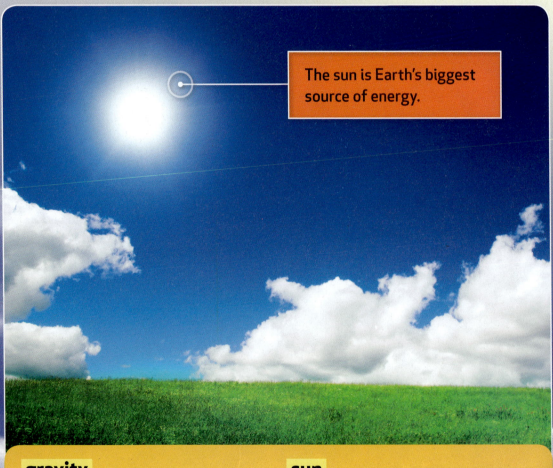

The sun is Earth's biggest source of energy.

gravity
Gravity is a force that pulls objects toward each other.

sun
The **sun** is the star that is nearest to Earth.

TECHTREK
myNGconnect.com

Student
eEdition

Digital
Library

Blue Sky Sunlight is white <mark>light</mark>. White light actually contains light of many different colors. Particles in Earth's air are just the right size to scatter the blue part of sunlight. When sunlight reaches the air, the particles bounce the blue light around. Blue light gets scattered throughout the sky. All the other colors continue on to Earth. That's why the sky looks blue to us.

WHY THE SKY IS BLUE

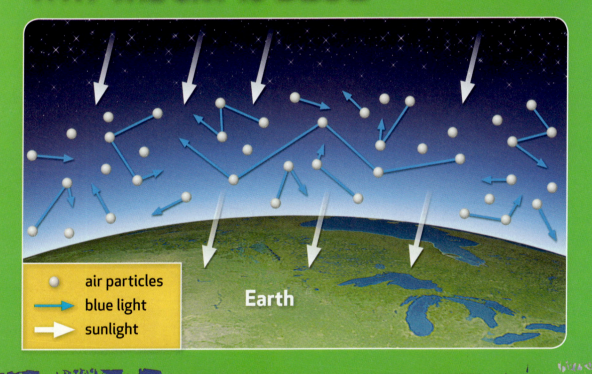

- ● air particles
- → blue light
- ⇨ sunlight

Earth

energy
Energy is the ability to do work or cause a change.

light
Light is a kind of energy you can see.

Rainbows Rainbows form when sunlight bends. Look at the picture of a prism. Remember, white light is made up of many colors. When white sunlight passes through the prism, the different colors separate into a rainbow.

Light bends when it hits the prism.

Light leaves the prism as separate colors.

White sunlight passes through a prism.

The same thing happens in the sky. Just after it rains, tiny drops of water are still in the air. The raindrops act like tiny prisms. White sunlight passes through raindrops. The different colors of light bend differently through the raindrops and separate. The colors form part of a circle through the sky, or a rainbow.

The colors in a rainbow always follow the same pattern: red, orange, yellow, green, blue, indigo, violet.

Halos When a storm is coming, sometimes high, wispy clouds move in first. The <mark>temperature</mark> of the air is so cold that the clouds are made of ice. These tiny ice pieces bend the sunlight. This bending forms a rainbow in the sky that looks like a circle around the sun.

Have you ever seen a ring of light around the sun? The ring is really in Earth's air, not around the sun.

temperature

Temperature is a measure of how hot or cold something is.

Sun Dogs Sometimes a full circle of light doesn't form, like it does with a halo. The ice crystals in the clouds bend sunlight in a different way. Then only small parts of a circle of light form. These parts form on either side of the sun and are called sun dogs. Sun dogs often happen when the sun is low in the sky.

Sometimes a sun dog appears on only one side of the sun. At other times a sun dog appears on both sides.

Auroras One of nature's most colorful light shows is called an aurora. Auroras often look like glowing curtains of light. Auroras happen when the sun's extreme heat produces tiny superhot particles. These particles shoot out into space. They crash into particles in Earth's air near the North and South poles. These crashes release energy that **transforms** into different colors of light.

TECHTREK
myNGconnect.com

Digital Library

Most auroras are green, red, purple, or white.

transform

to **transform** is to change.

Unlike most solar light shows, auroras happen at night. Other shows, like rainbows, halos, or sun dogs happen during the day. That's when the sun is out. Whether they happen during the day or at night, these amazing shows all have one thing in common. The sun's energy produces them all. The only thing you need to do to see them is look up!

This view of an aurora was taken by one of the crew on a space shuttle.

CHAPTER 4

SHARE AND COMPARE

Turn and Talk How does sunlight produce different effects in the sky? Form a complete answer to this question together with a partner.

Read Select two pages in this section. Practice reading the pages. Then read them aloud to a partner. Talk about why the pages are interesting.

Write Write a conclusion that tells the important ideas about solar light shows. State what you think is the Big Idea of this section. Share what you wrote with a classmate. Compare your conclusions.

Draw Form groups of five. Have each person draw a different solar light show. Each person should label their drawings and write captions to explain the cause of the light show. Present your drawings and explanations to the class.

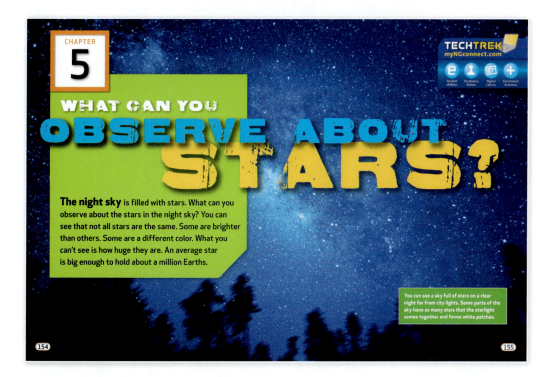

CHAPTER

5

WHAT CAN YOU

OBSERVE ABOUT
STARS?

TECHTREK
myNGconnect.com

Student   Vocabulary   Digital   Enrichment
edition   Games   Library   Activities

The night sky is filled with stars. What can you observe about the stars in the night sky? You can see that not all stars are the same. Some are brighter than others. Some are a different color. What you can't see is how huge they are. An average star is big enough to hold about a million Earths.

You can see a sky full of stars on a clear night far from city lights. Some parts of the sky have so many stars that the starlight comes together and forms white patches.

154

155

In Chapter 5, you will learn:

FLORIDA NEXT GENERATION SUNSHINE STATE STANDARDS

SC.3.E.5.1 Explain that stars can be different: some are smaller, some are larger, and some appear brighter than others; all except the Sun are so far away that they look like points of light. **THE NIGHT SKY, PROPERTIES OF STARS, OBSERVING STARS**

SC.3.E.5.5 Investigate that the number of stars that can be seen through telescopes is dramatically greater than those seen by the unaided eye. **OBSERVING STARS**

SC.3.E.5.5 **Science in a Snap!** Investigate that the number of stars that can be seen through telescopes is dramatically greater than those seen by the unaided eye.

WHAT CAN YOU OBSERVE ST

The night sky is filled with stars. What can you observe about the stars in the night sky? You can see that not all stars are the same. Some are brighter than others. Some are a different color. What you can't see is how huge they are. An average star is big enough to hold about a million Earths.

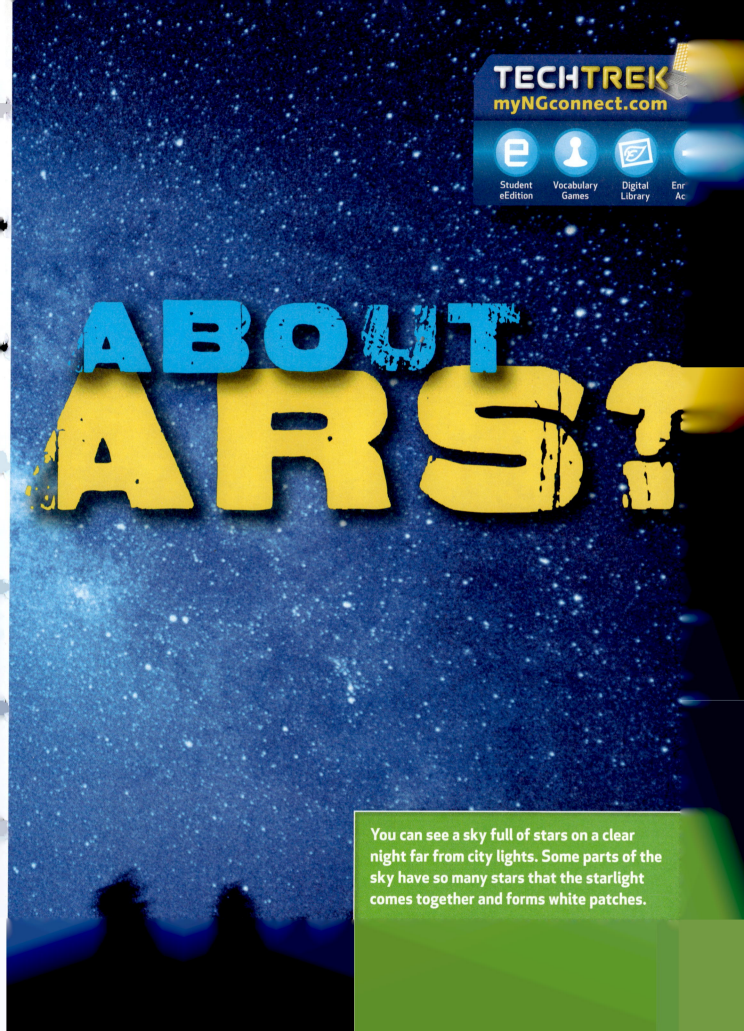

TECHTREK
myNGconnect.com

Student
eEdition

Vocabulary
Games

Digital
Library

Enr
Ac

ABOUT STARS?

You can see a sky full of stars on a clear night far from city lights. Some parts of the sky have so many stars that the starlight comes together and forms white patches.

SCIENCE VOCABULARY

star (STAR)

A **star** is a glowing ball of hot gases. (p. 158)

Stars look like points of light because they are so far away from Earth.

property (PROP-ur-tē)

A **property** is something about an object that you can observe. (p. 160)

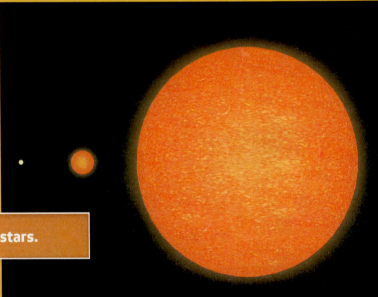

Size is a property of stars.

my
Science Vocabulary

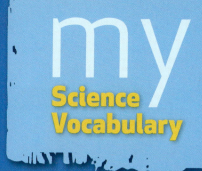

brightness (BRĪT-nes)	**star** (STAR)
property (PROP-ur-tē)	**telescope** (TEL-uh-scōp)

TECHTREK
myNGconnect.com

Vocabulary
Games

brightness (BRĪT-nes)

Brightness is the amount of light that reaches your eye from an object such as a star. (p. 160)

The brightness of a light depends partly on its distance from the viewer.

telescope (TEL-uh-scōp)

A **telescope** is a tool that magnifies objects and makes them look closer and bigger. (p. 166)

The telescopes of today are much more powerful than early telescopes.

The Night Sky

Away from bright city lights, you can see many stars in the night sky. A star is a glowing ball of hot gases. Stars are so far away that they look like small points of light. In a city, you might see only the brightest stars at night. This is because streetlights and building lights wash out most of the starlight. So most stars are hard to see.

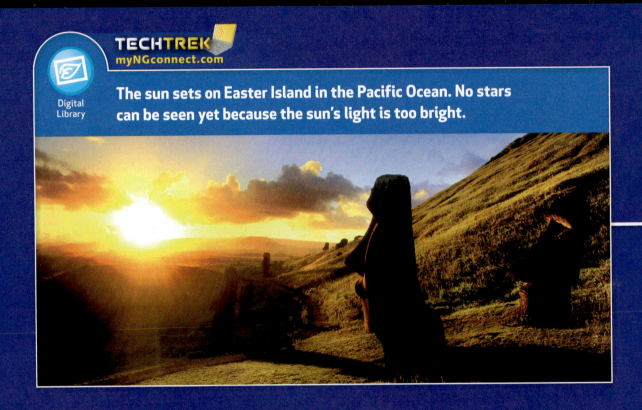

TECHTREK
myNGconnect.com

Digital Library

The sun sets on Easter Island in the Pacific Ocean. No stars can be seen yet because the sun's light is too bright.

The sun is a star. It is closer to Earth, so it looks much larger than other stars. You can see the sun only during the daytime. This is because during the day, your part of Earth faces the sun. At night, your part of Earth turns away from the sun. You can see the other stars in the sky then, because light from the sun no longer washes out the light from other stars.

As night begins to fall you can begin to see stars.

With no sunlight, the night sky is full of stars.

Before You Move On

1. What is a star?
2. Do you think it used to be easier for people to see the stars at night? Why or why not?
3. **Infer** The planet Jupiter is farther away from the sun than Earth. What do you think the sun would look like from Jupiter?

Brightness and Size Scientists classify, or group, stars by their **properties**. A property is something about an object that you can observe with your senses. Two properties of stars are **brightness** and size. Brightness is the amount of light that reaches your eye from an object such as a star.

Some stars look brighter than others because they are closer to Earth. It's like a line of streetlights. The lights are all the same size and give off the same amount of light. But the closer ones look brighter.

Why do some of these streetlights look brighter than others?

Unlike the streetlights, all stars are not the same size. Some look brighter because they are larger. In fact, some stars are called giants or supergiants. Others are much smaller and are called dwarf stars.

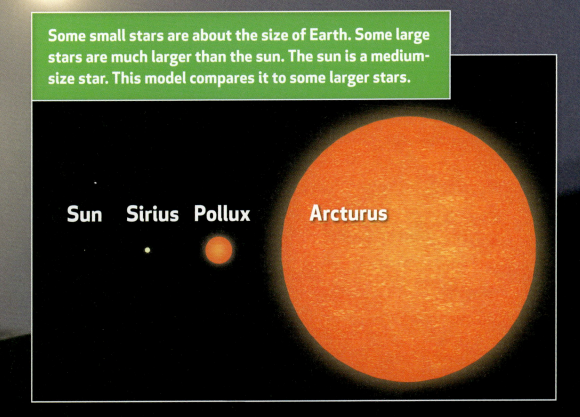

Some small stars are about the size of Earth. Some large stars are much larger than the sun. The sun is a medium-size star. This model compares it to some larger stars.

Sun   Sirius   Pollux   Arcturus

Temperature Size plays a part in how bright a star is. But the temperature of the star is important too. Temperature is how hot or cold something is. All stars are hot, but some are hotter than others. How can you tell which stars are hotter? It's all about color. Look at the metal pot in the picture. The pot is black. As it heats up, it changes color.

TECHTREK
myNGconnect.com

Enrichment Activities

As the bottom of the pot heats up, it changes color. The temperature shows you how hot the pot is. If it could keep getting hotter, the whole pot would glow blue.

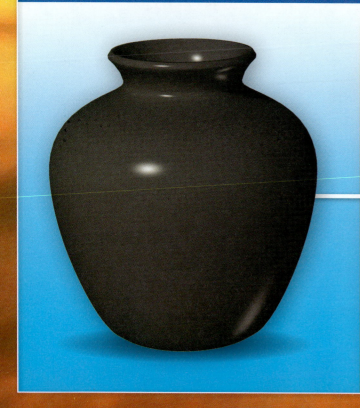
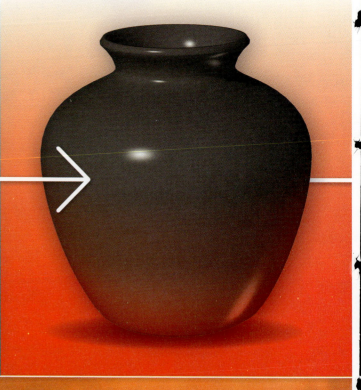

At first, the black pot glows red. As it gets hotter and hotter, the bottom of the pot changes color. It goes from red to orange to yellow to white to blue. Objects that glow red are cooler than those that glow white or blue.

Color At first glance, you may think all stars are white or light yellow. But look closely. Stars are different colors. And like the metal pot, a star's color is a clue to how hot it is. The coolest stars are red. The hottest stars are blue. White or blue stars are brighter because they are hotter and give off the most energy for their size.

Which star is the coolest in this photo?

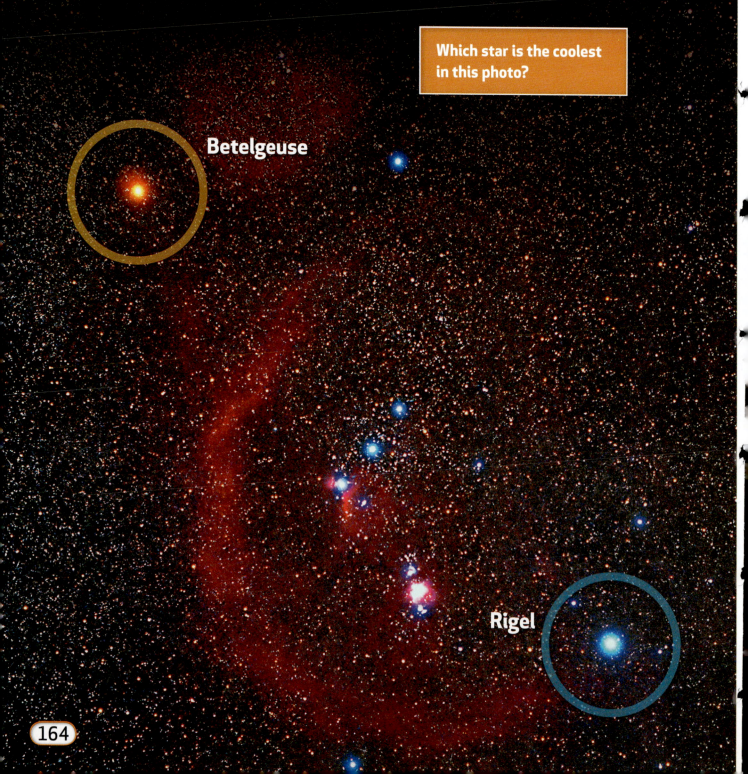

Betelgeuse

Rigel

Betelgeuse is one of the biggest stars in the sky. But another large star, Rigel, is brighter because its temperature is hotter. Look at the stars shown in the chart. Compare the color and temperature of the sun with the other stars in the chart.

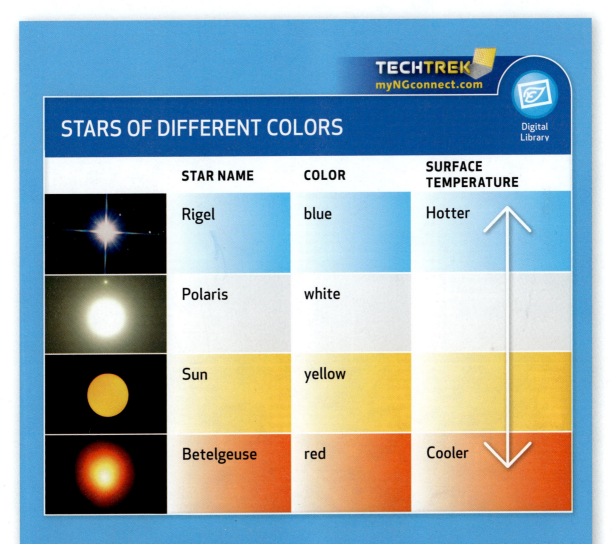

TECHTREK
myNGconnect.com

Digital Library

STARS OF DIFFERENT COLORS

	STAR NAME	COLOR	SURFACE TEMPERATURE
	Rigel	blue	Hotter
	Polaris	white	
	Sun	yellow	
	Betelgeuse	red	Cooler

Before You Move On

1. What two properties play a role in how bright a star is?
2. What is the difference between red stars and blue stars?
3. **Evaluate** If a star looks brighter than another star, does that always mean it is bigger than the other star? Why or why not?

Observing Stars

Scientists have used **telescopes** for hundreds of years. A telescope is a tool that magnifies objects. It makes faraway objects look closer and bigger. You can see more detail and learn more about the objects. A telescope collects more light than your eyes can. So with a telescope, you can see more stars than with your eyes alone.

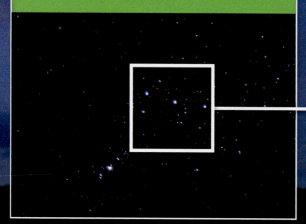

Here are some stars you might see with your eyes alone. Notice the three stars that line up.

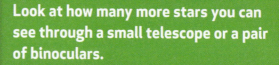

Look at how many more stars you can see through a small telescope or a pair of binoculars.

Early telescopes were not very powerful. But telescopes today are bigger and much more powerful. Scientists use them to see deeper and deeper into space. They have discovered more stars than ever before. In the future, new powerful telescopes will help scientists learn even more about stars.

The stars can be seen as night falls over Caspian Lake, Vermont.

At the Gemini North Observatory in Hawaii, scientists use a large telescope to observe the night sky. They use the telescope, computers, and other tools to watch and take pictures of stars. Gemini North was built on top of a mountain so scientists can see the whole sky.

With a powerful telescope, scientists can see the colorful light from gas and dust in space.

The Gemini North Observatory sits on top of a mountain. The observatory is 4200 meters (14,000 feet) above sea level. That's higher than most clouds.

Using Gemini North's telescope, scientists watch many different stars. They observe stars that have just formed. They observe stars that have faded away or exploded.

TELESCOPE

This telescope is inside the Gemini North Observatory. Special mirrors focus and reflect light.

A. The first mirror is at the base of the telescope. It is over 8 meters (27 feet) wide.

B. Light from the first mirror is reflected up to the second mirror.

A camera inside the telescope records images.

Before You Move On

1. What does a telescope do?
2. What might you see through a telescope that you cannot see with just your eyes?
3. **Infer** Why would the Gemini North Observatory not be as useful if it were built in a city?

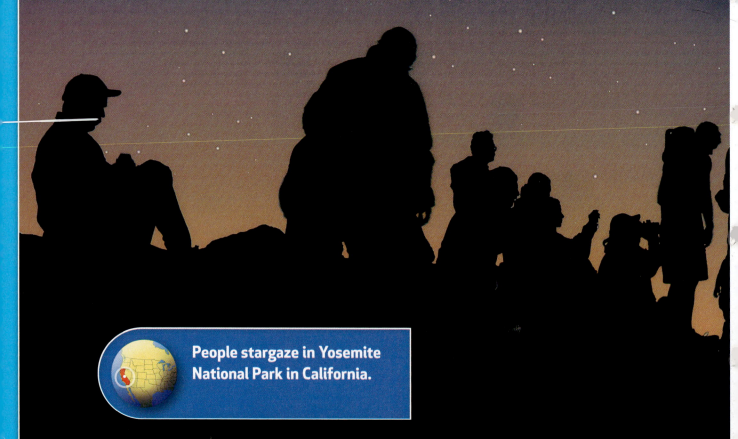

NATIONAL GEOGRAPHIC

ANCIENT LIGHT

Proxima Centauri is Earth's second-closest star. It is four light-years away. So if this star exploded today, we wouldn't see the explosion for another four years! How can that be? You see an object only because light travels from the object to your eyes. Stars are so far away that their light takes years to reach us. Scientists use a unit called a light-year to measure how far stars are from Earth. A light-year is the distance light moves in a year. A light-year is about 9.5 trillion kilometers (about 6 trillion miles).

People stargaze in Yosemite National Park in California.

Sunlight takes only eight minutes to reach Earth. When explosions happen on the sun, we see them eight minutes later. Compare eight minutes to 800 years—that's how long it takes light from a star like Rigel to reach Earth. Rigel is 800 light-years away. Most stars are thousands or millions of light-years away.

This disc-shaped group of stars is almost 3 million light-years away. If scientists would observe something happening there, the event would have happened 3 million years ago.

This picture was taken over the rocks of Joshua Tree National Park in California. It shows the blue-white star Sirius. It is the brightest star seen in the night sky. It is nearly 9 light years away.

Conclusion

Stars are huge balls of hot gases. They are so far away that they look like points of light in the night sky. Stars are not all the same. Some are closer to Earth than others. Some are bigger and brighter than others. Stars also have different surface temperatures, which make them different colors. A star's brightness depends on its size and surface temperature. You can see many more stars with a telescope than you can with your eyes alone.

Big Idea You can observe several properties of stars including brightness, size, and color.

PROPERTIES OF STARS

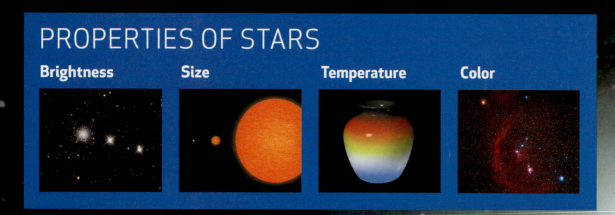

Brightness   Size   Temperature   Color

Vocabulary Review

Match the following terms with the correct definition.

A. star

B. brightness

C. telescope

D. property

1. A glowing ball of hot gases
2. A tool that magnifies objects and makes them look closer and bigger
3. The amount of light that reaches your eye from an object such as a star
4. Something about an object that you can observe

Big Idea Review

1. **Organize** Order the star colors from coolest to hottest: blue, red, white, yellow.

2. **Describe** How does a telescope work?

3. **Explain** Why do stars seem so small when you look at them with your eyes alone?

4. **Compare and Contrast** How are hot stars and cool stars alike and different?

5. **Make Judgements** Why do you think it is better to build an observatory high on a mountain rather than lower down?

6. **Generalize** You have learned about the properties of stars. Use these properties to describe the sun.

Write About the Night Sky

Cause and Effect The sun is a star you can see during the day. Write a few sentences to explain why you can't see other stars during the day.

CHAPTER 5

EARTH SCIENCE EXPERT: ASTRONOMER

Want to study the stars and beyond? Astronomer Jason Kalirai talks about outer space.

What does an astronomer do?

I use large telescopes to take pictures of planets, stars, and galaxies in the universe. I use telescopes on Earth and in space. I then study the pictures to learn more about these objects.

What's your favorite part of your job?

My favorite part is when I first look at a new picture of a star or galaxy. It's exciting to see something for the first time, especially if it's unexpected.

Jason Kalirai uses computers to study pictures of stars.

How did you become interested in astronomy?

As a kid, I loved looking at the night sky. I learned all I could about the stars and planets.

What's a typical day like for you?

I usually go to work from 9:00 a.m. to 5:00 p.m., like most people. At work, I spend a lot of time talking to fellow astronomers. We share and discuss our observations. I also spend a lot of time traveling. I've been all over the U.S., including Hawaii. I've been to other countries too, including Australia, New Zealand, India, Japan, Egypt, and many places in Europe. While visiting these places, I get new observations and meet new astronomers.

Why should students become astronomers?

I find my job to be the coolest one in the world, and I want others to see why. We need new, young scientists who will think creatively. That means asking questions no one has asked before. It means observing something and explaining it in new ways.

Jason zeroes in on a group of stars to learn more about them.

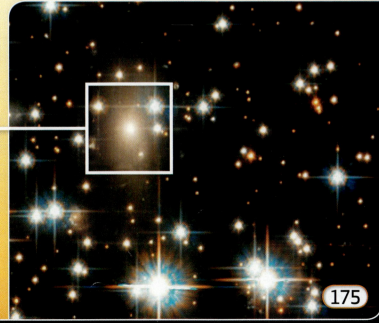

BECOME AN EXPERT

The Hubble Space Telescope:
Super Stargazer

From Earth, it is hard to see many of the glowing balls of hot gases called **stars**. This is because the moving gases in the air around Earth can block starlight. To get a better view of stars and space, scientists launched the Hubble Space **Telescope** in 1990.

The Hubble Space Telescope flies 569 kilometers (353 miles) above Earth.

stars

A **star** is a glowing ball of hot gases.

telescopes

A **telescope** is a tool that collects light from objects and magnifies them.

The space telescope flies high above Earth. It circles our planet once every 97 minutes. All the while, it points into space and takes pictures. The telescope sends these pictures back to Earth. Studying the photos has taught scientists a lot about the universe. Some of Hubble's photos show colorful clouds of gas and dust in space. Such a cloud is called a nebula. Stars are born in a nebula.

Stars are forming in this cloud of gas and dust called the Eagle Nebula.

A Star is Born

Stars form when bits of gas and dust come together inside a nebula. A force called gravity pulls together gas and dust. This material forms a clump, which grows larger and larger. As the clump grows, so does its gravity.

TECHTREK
myNGconnect.com

Digital Library

Many stars form in the Swan Nebula. This nebula is 5,000 light-years from Earth.

Gravity pulls more and more gas and dust into the clump. It also pulls the clump into a tight ball. As the ball gets tighter and tighter, its **properties** change. The temperature rises. If the ball becomes hot enough, it starts shining. At that point, a star is born. Its **brightness** might be seen from Earth. All of this takes millions of years.

HOW DO STARS FORM?

Gravity begins to pull gas and dust together into a clump.	The clump forms a ball called a protostar.	Gas and dust in the protostar heat up enough to give off a lot of energy. A star is born.

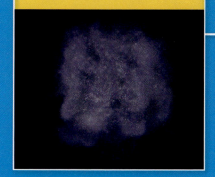 → 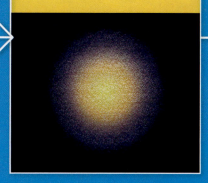 →

property

A **property** is something about an object that you can observe.

brightness

Brightness is the amount of light that reaches your eye from an object such as a star.

Dying Stars

Photographs from the Hubble Space Telescope are also helping scientists learn about the end of a star's life. No star lasts forever, and that includes our sun. It is five billion years old. But don't worry. It should stay as it is for another five billion years. Then it will start to cool down, turn red, and grow larger.

The supergiant red star in the center of this photo is dying. Light from this star let Hubble see the swirling nebula around it. The other stars are at different distances. They are behind and in front of the nebula.

Near the end of its life, the sun might even grow large enough to touch Earth! A star that size is called a red giant. After a while, the sun will stop growing and shed its outer layers. Photos from Hubble have shown that these layers float away gently, like a puff of smoke. They form a nebula around the dying star. Later, it will fade to black and die. But new stars will form in the nebula left behind.

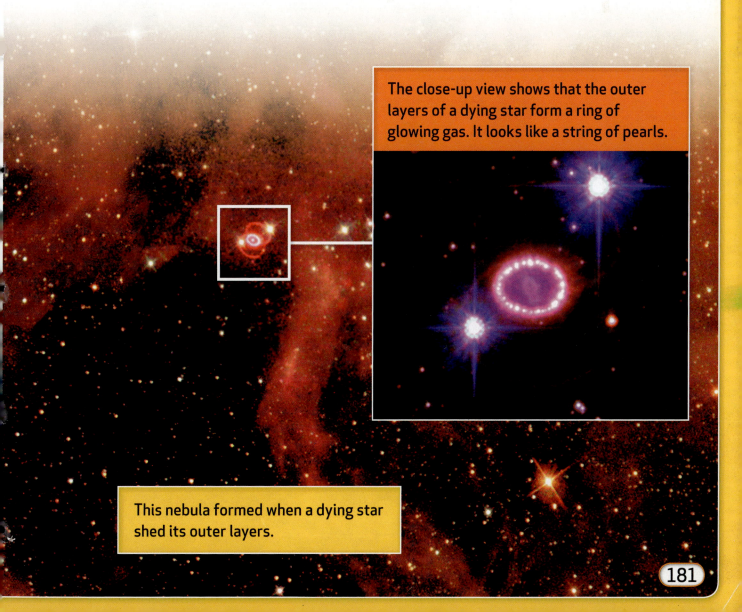

The close-up view shows that the outer layers of a dying star form a ring of glowing gas. It looks like a string of pearls.

This nebula formed when a dying star shed its outer layers.

Looking Ahead

The Hubble Space Telescope has taught us about the lives of stars and much more. But the Hubble can only do so much. Soon a new telescope will join the Hubble high above Earth. In 2014, scientists plan to launch the James Webb Space Telescope. The Webb telescope will let scientists look even deeper into space.

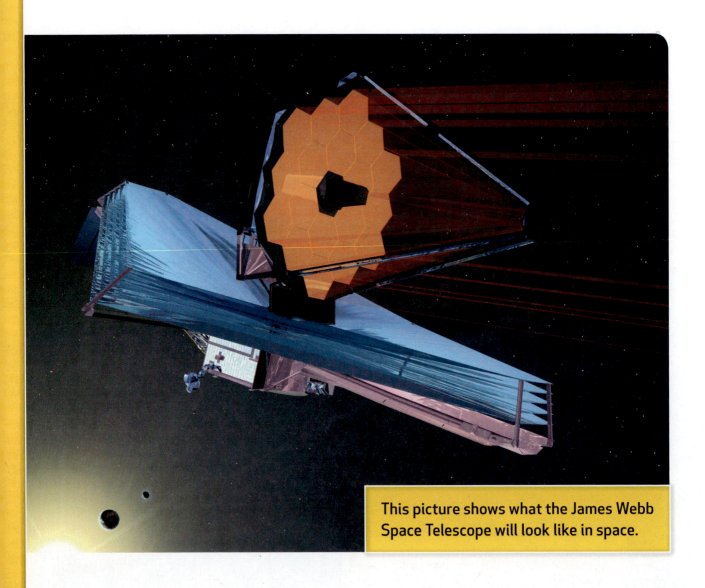

This picture shows what the James Webb Space Telescope will look like in space.

For 20 years, the Hubble Space Telescope has taken pictures of faraway stars and galaxies. Scientists have studied photos taken by Hubble to learn about the life cycle of stars. In the future, Hubble and the James Webb Space Telescope will tell us even more about our amazing universe.

In the Orion Nebula, several white-hot stars have already formed. The swirling clouds of gas and dust will form even more stars.

CHAPTER 5

SHARE AND COMPARE

Turn and Talk   How are stars born and how do they die? Form a complete answer to this question together with a partner.

Read   Select two pages in this section. Practice reading the pages. Then read them aloud to a partner. Talk about why the pages are interesting.

Write   Write a conclusion that tells the important ideas you have learned about the Hubble Space Telescope. State what you think is the Big Idea of this section. Share what you wrote with a classmate. Compare your conclusions.

Draw   Form groups of three. Have each person draw a star at different points in its life. Each person should label their drawings and write captions to explain what is happening in the drawings. Present your drawings and explanations to the class.

FLORIDA
PHYSICAL
SCIENCE

What Is Physical Science?

Physical science is the study of the physical world around you. This type of science investigates the properties of different objects, as well as how those objects interact with each other. Physical science includes the study of matter, motion and forces, and many kinds of energy, including light and electricity. People who study how all of these things work together are called physical scientists.

You will learn about these aspects of physical science in this unit:

HOW CAN YOU DESCRIBE AND MEASURE MATTER?

Matter is anything that has mass and takes up space. Physical scientists study all of the different properties of matter. These include size, shape, color, texture, and hardness, as well as mass and volume.

HOW DOES WATER CHANGE?

Physical scientists study the many different properties of water. They also study how temperature changes can cause water to change from one state of matter to another.

WHAT IS ENERGY?

Physical scientists study energy in all of its forms. They also learn about how energy can cause changes in the physical world. The different kinds of energy include light, sound, electrical, and mechanical.

WHAT IS LIGHT?

Light is a kind of energy that you can see. Physical scientists study the properties of light. They also learn about how light can change direction and even cause objects to heat up.

MEET A SCIENTIST

Constance Adams: Space Architect

Constance Adams is a space architect with NASA and a National Geographic Emerging Explorer. One of her first projects with NASA was TransHab, designed to be a transit habitat for the first human mission to Mars.

Constance and her team of experts bring together innovations from diverse disciplines such as architecture, engineering, industrial design, and sociology with the hopes of solving complex design issues. "When you have a brand-new problem, you need as many tools as you can get. Who knows—an approach from a very different field might give you the insight you need. For example, I'm working to forge communication between advanced engineering and consumer-product design to bring more user-centered designs to aerospace," says Constance.

As a space architect, Constance works with space agencies, including NASA, to design and build components for spacecraft and the International Space Station.

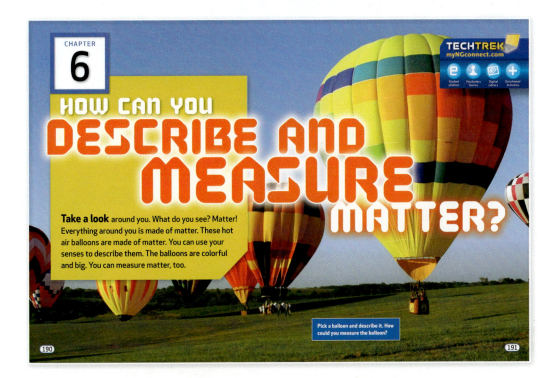

CHAPTER
6

HOW CAN YOU
DESCRIBE AND
MEASURE
MATTER?

Take a look around you. What do you see? Matter! Everything around you is made of matter. These hot air balloons are made of matter. You can use your senses to describe them. The balloons are colorful and big. You can measure matter, too.

TECHTREK
myNGconnect.com

Pick a balloon and describe it. How could you measure the balloon?

190

191

In Chapter 6, you will learn:

FL RIDA NEXT GENERATION SUNSHINE STATE STANDARDS

SC.3.P.8.2 Measure and compare the mass and volume of solids and liquids.
MEASURING MASS, MEASURING VOLUME

SC.3.P.8.3 Compare materials and objects according to properties such as size, shape, color, texture, and hardness. **PROPERTIES OF MATTER**

SC.3.P.8.3 Science in a Snap! Compare materials and objects according to properties such as size, shape, color, texture, and hardness.

CHAPTER

6

HOW CAN YOU DESCRIBE MEA

Take a look around you. What do you see? Matter! Everything around you is made of matter. These hot air balloons are made of matter. You can use your senses to describe them. The balloons are colorful and big. You can measure matter, too.

TECH**TREK**
myNGconnect.com

Student
eEdition

Vocabulary
Games

Digital
Library

Enrichment
Activities

AND SURE MATTER?

Pick a balloon and describe it. How could you measure the balloon?

SCIENCE VOCABULARY

matter (MA-ter)

Matter is anything that has mass and takes up space. (p. 194)

The airplane and buildings are made of matter.

texture (TEKS-chur)

Texture describes the surface of any area made up of matter. (p. 198)

The metal armrest has a smooth texture.

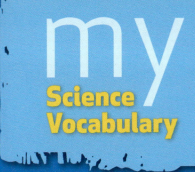

my
Science Vocabulary

mass (MAS)	**texture** (TEKS-chur)
matter (MA-ter)	**volume** (VOL-yum)

TECHTREK
myNGconnect.com

Vocabulary
Games

mass (MAS)

Mass is the amount of matter in an object. (p. 200)

The item that hangs lower has a greater mass.

volume (VOL-yum)

Volume is the amount of space matter takes up. (p. 202)

The city checks the volume in the tank to be sure there is enough fresh water.

Properties of Matter

What do you see in this photo? Everything you see is made of  **matter** . Matter takes up space and has mass.

You can describe the airplane by telling about its properties. Properties are things you observe with your senses. The airplane, for example, is big, blue, and white.

Size Size is a property of matter. How long do you think this airplane is? You could measure it to find out.

Measuring the size of matter can be important. The runway in the photo needs to be long enough for the airplane to safely land. You could measure both the airplane and runway to find out how long they are.

Choose two buildings and compare their sizes. Which is taller? Which is wider?

Color and Shape When you leave an airplane, you have to find your suitcase. How do you find it? You probably observe the properties of color and shape. These suitcases are different shapes and colors. Which do you think would be easiest to find?

TECHTREK
myNGconnect.com

Digital Library

Color is a property you observe with your sense of sight.

Shine a flashlight onto a white piece of paper. Then cover the flashlight with a piece of colored cellophane.

What happens to the color of the light?

Shine the flashlight on a white piece of paper.

Shape is important! The round wheels on these suitcases help the family move quickly through the airport. Square wheels wouldn't work.

Texture and Hardness Airplane passengers want comfortable seats. That's why the chairs are covered with smooth fabric like cotton instead of scratchy fabric like wool. *Smooth* and *scratchy* are words that describe **texture** . Texture is a property of an object's surface you feel by touching.

If you look at cotton and wool from a distance, they look very similar. If you look at them up close with a magnifying glass, though, the wool looks bumpier than the cotton.

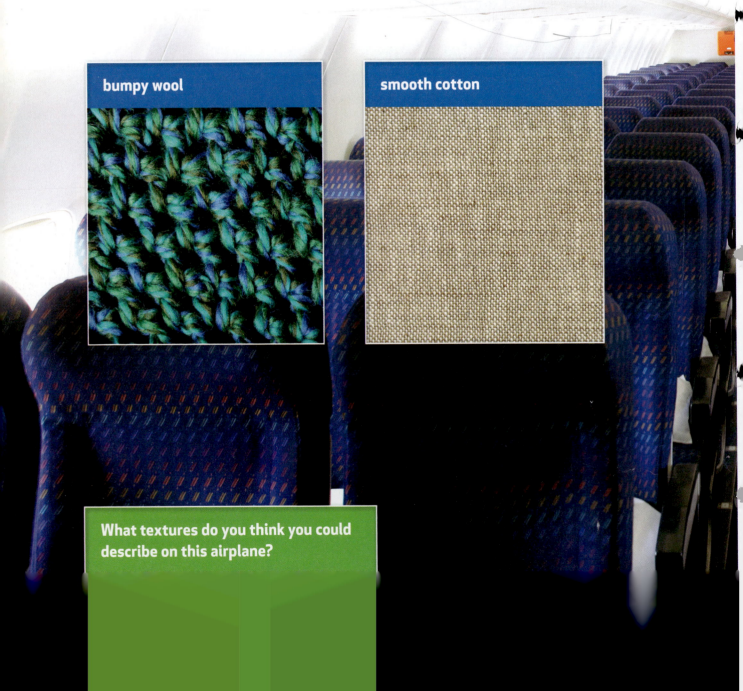

bumpy wool

smooth cotton

What textures do you think you could describe on this airplane?

If you pushed your finger against an airplane seat, your finger would make a dent. If you pushed your finger against the outside of an airplane, your finger could not press into the surface. The outside of an airplane is harder than an airplane seat. Hardness is another property that describes matter.

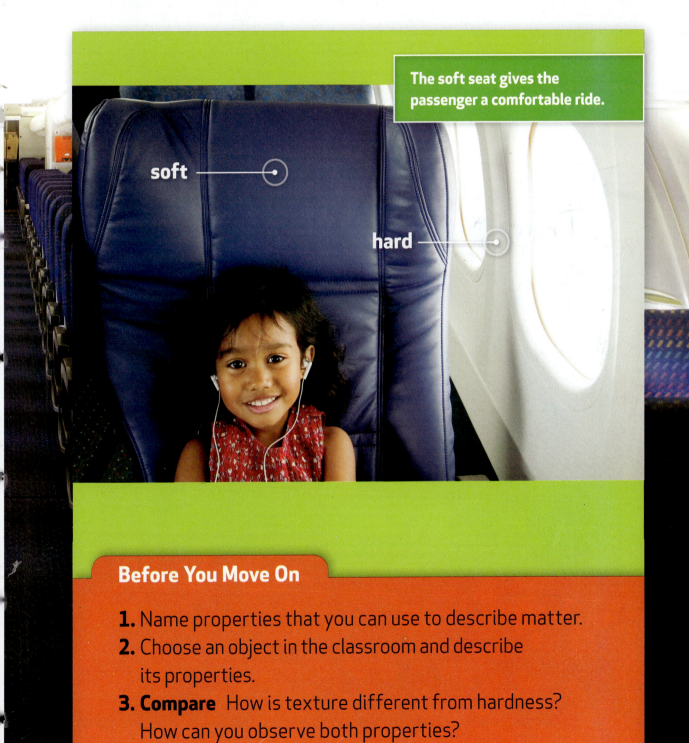

The soft seat gives the passenger a comfortable ride.

soft

hard

Before You Move On

1. Name properties that you can use to describe matter.
2. Choose an object in the classroom and describe its properties.
3. **Compare** How is texture different from hardness? How can you observe both properties?

Measuring Mass

You can compare sizes of objects by measuring their **mass** . Mass is the amount of matter in an object. Mass is measured in grams.

A balance can help you compare masses. The object that hangs lower has a greater mass. Do you measure your weight on a scale? Mass and weight are not the same thing. An object weighs more when there is more gravity. Earth has more gravity than the moon, so you would weigh more on Earth. You would have the same amount of matter, though, in both places.

TECHTREK
myNGconnect.com

Digital Library

Which has a greater mass—the stones or the feathers? Explain how you know.

You can use a balance to measure the mass of a solid. How do you measure the mass of a liquid? First, measure the mass of an empty container. Add the liquid to the container. Then, measure the mass of the container and the liquid. Subtract the first measurement from the second measurement. The remainder is the mass of the liquid.

It is important to measure the mass of the container before adding the liquid.

Measure the mass of the container and liquid. What should you do next?

Before You Move On

1. Identify the unit for measuring the mass of an object.
2. Describe how to measure the mass of a liquid.
3. **Compare** How does your mass on Earth compare to your mass on the moon? How does your weight on Earth compare with your weight on the moon?

Measuring Volume

Which holds more water, the water bottle or the water tank? The tank holds more water, but how could you figure out exactly how much more? You could compare **volume** . Volume is a property of matter. The amount of space that matter takes up is its volume. Liquid volume is measured in liters or gallons.

TECHTREK
myNGconnect.com

Digital Library

The volume of the bottle is one liter. The volume of the tank is many thousands of liters.

Look at the two containers below. Do you think they are holding the same volume of liquid? The volume of liquid in one container looks greater than the volume in the other container. Both containers hold the same amount of liquid, though. The shape of a container can fool our eyes into thinking there is more liquid in one container than in another.

You can use different sized beakers to measure the volumes of liquids.

You can use math to figure out the volume of a solid. First, measure the solid's length (l). Length means how long something is. Measure its width (w). Width means how wide something is. Measure its height (h). Height means how tall something is. Then, multiply length by width by height. The result is the volume of the solid.

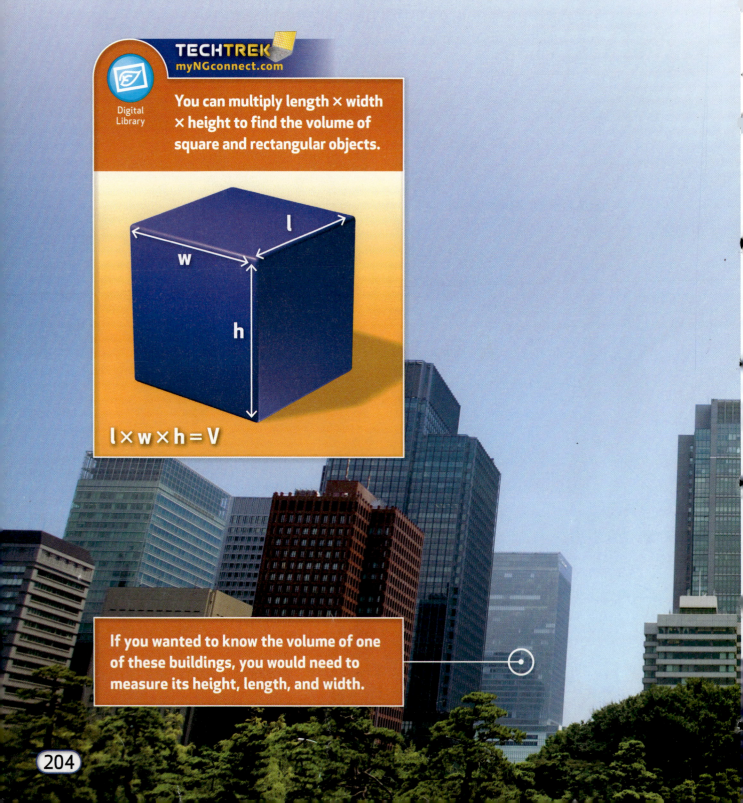

TECHTREK
myNGconnect.com

Digital Library

You can multiply length × width × height to find the volume of square and rectangular objects.

l × w × h = V

If you wanted to know the volume of one of these buildings, you would need to measure its height, length, and width.

If you want to find the volume of a solid block, you can measure the length, width, and height. What if you want to find the volume of an object with an odd shape, like a rock? Put water in a beaker. Measure its volume. Then, put the object in the beaker. Measure the volume again to see how far the water rose. The difference in the water's volume is the volume of the object.

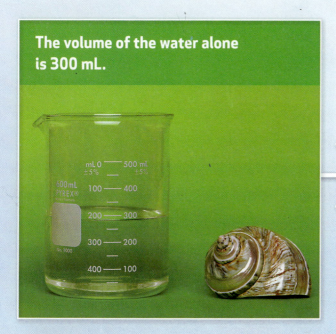

The volume of the water alone is 300 mL.

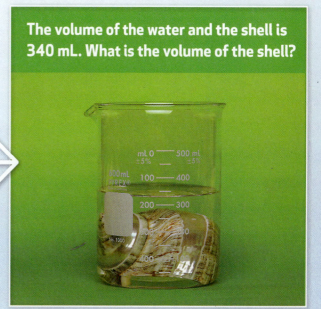

The volume of the water and the shell is 340 mL. What is the volume of the shell?

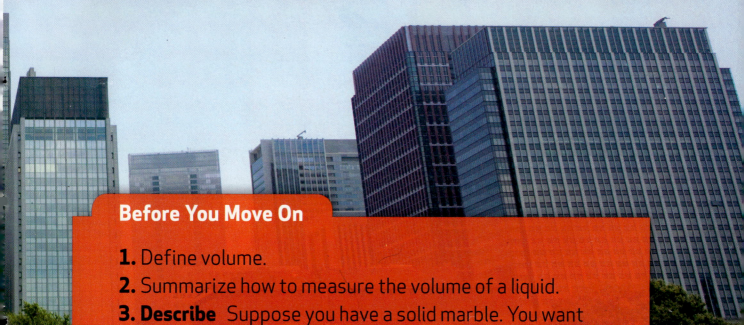

Before You Move On

1. Define volume.
2. Summarize how to measure the volume of a liquid.
3. **Describe** Suppose you have a solid marble. You want to measure its volume. Describe what you would do.

GIANTS OF THE PUMPKIN PATCH

You can describe and measure things that people make and things that you find in nature. How would you describe a pumpkin that weighs almost as much as a small car? You might describe it as amazing or gigantic. Scientists, however, would focus on its shape, color, texture, and mass. The mass of one of the largest pumpkins on record is 768 kilograms. That's a whopping 1,689 pounds!

How would you describe these pumpkins using the properties of size, color, and texture?

Small pumpkins often have round shapes. Giant pumpkins may have lumpy shapes. Giant pumpkins grow in different colors from orange to white. They are hard, and they usually have smooth textures, at least on the outside. The texture of the material on the inside is stringy.

You might think that a pumpkin that has the same mass as an adult grizzly bear would make a lot of pumpkin pies. The pies would not taste good, though. People who grow giant pumpkins don't grow them for food. Instead, they enter the pumpkins in contests. They just want to grow the largest pumpkins they can!

This pumpkin did not win a size prize because of a tiny crack in the bottom.

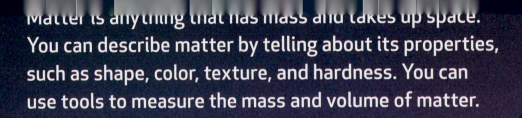

Matter is anything that has mass and takes up space. You can describe matter by telling about its properties, such as shape, color, texture, and hardness. You can use tools to measure the mass and volume of matter.

Big Idea Matter can be described and measured.

SOME PROPERTIES OF MATTER

Color	Size	Shape	Texture

Vocabulary Review

Match the following terms with the correct definition.

A. matter   **1.** The amount of matter in an object

B. mass   **2.** Describes the surface of any area made of matter

C. volume   **3.** Anything that has mass and takes up space

D. texture   **4.** The amount of space an object takes up

Big Idea Review

1. Define Tell what a property is. How can properties help you tell more about objects?

2. Explain Suppose you wanted to measure the volume of this book. How would you do it?

3. Summarize A friend asks why it's important to describe the properties of objects. How will you respond?

4. Compare and Contrast How are the texture of an apple and an orange different? What properties do the two objects have in common?

5. Infer You touch an object, and your finger makes a dent in it. What property of matter are you testing? How would you describe the object?

6. Analyze Suppose you had a pile of rocks and you wanted to sort them. What properties might the rocks have in common? What properties might be different?

Write About Matter

Explain Observe the object below. Describe its properties. Explain how you would measure its volume and mass.

PHYSICAL SCIENCE EXPERT:
PHYSICAL CHEMIST

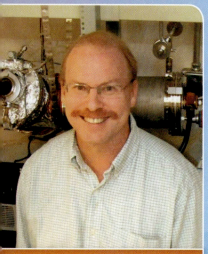

Rod Ruoff

Rod Ruoff is a physical chemist. He works at the University of Texas at Austin. He also started a company where he works with his team to research new ways of making energy. The team uses materials with interesting properties.

NG Science: What do you currently study?

Rod Ruoff: My team and I study properties of materials. We think about how these materials can make society better.

NG Science: When did you first know you wanted to be a physical chemist?

Rod Ruoff: Actually, I wanted to be either a professional hockey player or soccer player! In college, I took a physical chemistry course. I became fascinated with looking at different materials and thinking about new ways to use them.

NG Science: What type of research have you done?

Rod Ruoff: At first, I studied tiny pieces of matter to learn about how they stick together. Now I research materials made of carbon. Diamonds are made of carbon, and so is graphite. Graphite is the same material as the "lead" in your pencil. Individual layers of graphite are called "graphene" and we even make and study such atom-thick layers! I want to make materials that can help our environment.

TECHTREK
myNGconnect.com

Student
eEdition

Digital
Library

NG Science: Why is your research important?

Rod Ruoff: We think of new ideas that no one has ever thought of before. Our ideas can help others. We tackle hard problems.

NG Science: What is your favorite thing about being a physical chemist?

Rod Ruoff: There are many favorite things for me as a scientist. I think solving important problems, either alone or with a team, brings me the most joy.

TECHTREK
myNGconnect.com

Digital
Library

Rod Ruoff and a team member study the shape of matter.

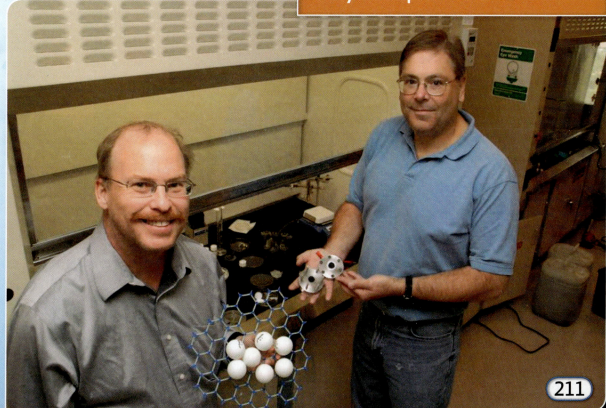

BECOME AN EXPERT

The Properties of Graphene

Graphene is an unusual type of **matter**. Graphene has something in common with diamonds and with the graphite that makes up your pencil lead. All these materials are a form of matter called carbon. Each form of carbon has different properties you can observe.

TECHTREK
myNGconnect.com

Digital
Library

FORMS OF CARBON

Graphene, graphite, and diamonds are different forms of carbon.

GRAPHENE	GRAPHITE	DIAMONDS
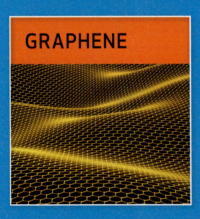	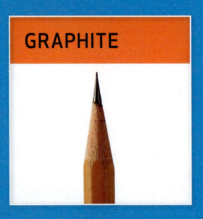	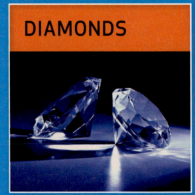

matter

Matter is anything that has mass and takes up space.

Color   Think about the different forms of carbon. What color are they? Natural carbon is very dark. So is the graphite in pencil lead. When you write, the dark graphite wears off the pencil and onto your paper. Diamonds have no color—they are clear. Like graphite, graphene is dark.

Graphite

Size Unlike diamonds and graphite, graphene is so thin, it's almost invisible! A sheet of graphene is less than one nanometer thick. Look at the photo of the ruler. Do you see how small 1 millimeter is? It takes 1 million nanometers to make 1 millimeter!

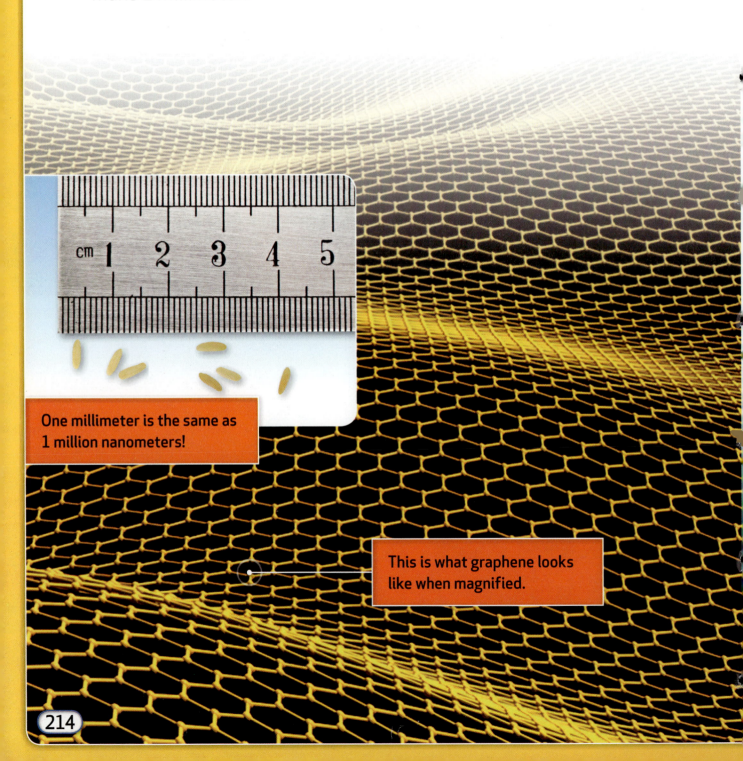

One millimeter is the same as 1 million nanometers!

This is what graphene looks like when magnified.

Even with its tiny size, graphene is incredibly strong. It's the strongest material ever tested. Think of a sheet of paper. Paper is not very strong. You can rip a sheet in half. Graphene is much thinner than paper. You might think you could rip a sheet of graphene in half, too. Graphene, though, is almost 200 times as strong as steel!

This steel beam is much stronger than this piece of paper.

Graphene is 200 times as strong as steel.

Hardness The graphite form of carbon is soft. The diamond form of carbon, though, is one of the hardest materials on Earth. Graphene is a hard material, too. Adding graphene to materials can make them stiff. If you hold a piece of paper on the side by its edges, it will bend toward the floor. A sheet of graphene is much thinner than a piece of paper. If you could hold a sheet of graphene, it would be stiff. Researchers are looking for ways to make materials stronger and stiffer by adding graphene to them.

Graphene is much thinner than this paper, but a sheet of graphene would not bend.

Shape and Texture

What makes graphene so strong? The property of shape is part of graphene's super strength. If you look at a very magnified view of graphene, you see it looks like chicken wire. Each of these shapes works together to form a strong network. The **texture** of a sheet of graphene is smooth.

Someone can see through chicken wire by looking through the holes.

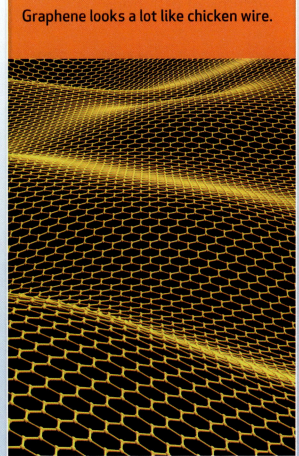

Graphene looks a lot like chicken wire.

texture

Texture describes the surface of any area made up of matter.

Mass and Volume Graphene does not have much **mass** . A layer large enough to cover a football field would have a mass of one gram, about the same mass as a teaspoon of sugar.

Do you remember how to measure **volume** ? What would be the volume of enough graphene to cover a football field? A football field is 110 meters (360 ft) long and 49 m (160 ft) wide. The height of graphene is less than a nanometer. So if you multiplied length × width × height, you'd find a very small volume of graphene.

MASS AND VOLUME OF GRAPHENE

How much graphene does it take to cover a football field?

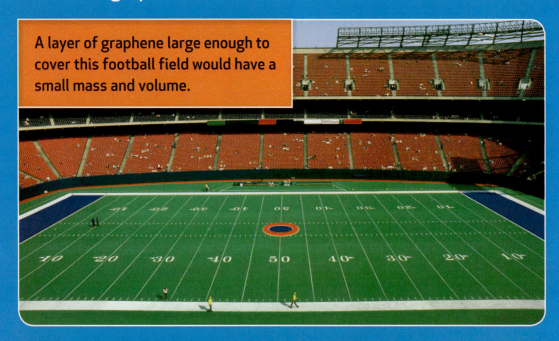

A layer of graphene large enough to cover this football field would have a small mass and volume.

mass

Mass is the amount of matter in an object.

volume

Volume is the amount of space matter takes up.

Graphene can store energy. It allows energy to flow very quickly, too. That means that graphene can be part of new technology. Scientists are excited about the many things they think graphene can do. Scientists think that graphene could help them make electronic paper like computer screens. The screens would be so thin, you could roll them up and carry them. Graphene might help scientists create faster computers and better cell phones. Graphene has an amazing future!

Graphene can be used in solar power cells. Graphene is so thin, it lets light through.

CHAPTER 6
SHARE AND COMPARE

Turn and Talk How can properties help us describe graphene? Form a complete answer to this question together with a partner.

Read Select two pages in this section. Practice reading the pages. Then read them aloud to a partner. Talk about why the pages are so interesting.

Write Write a conclusion that tells the important ideas you learned about graphene. State what you think is the Big Idea of this section. Share what you wrote with a classmate. Did your classmate write about the properties of graphene and what makes it unique?

Draw Suppose you magnified graphene even closer than the magnifications shown in this book. Draw what you think this close-up view would look like. Combine your drawings with those of your classmates to create a graphene gallery.

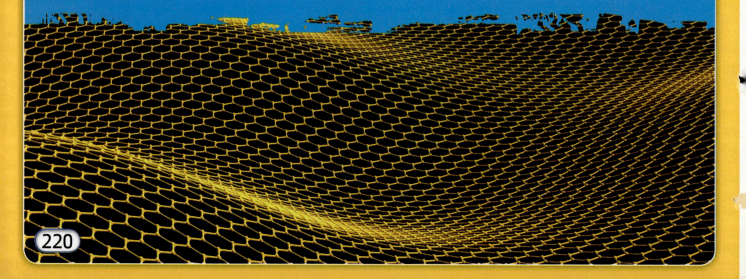

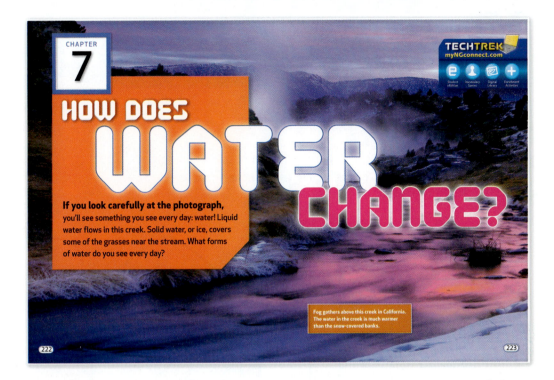

CHAPTER
7

HOW DOES WATER CHANGE?

If you look carefully at the photograph, you'll see something you see every day: water! Liquid water flows in this creek. Solid water, or ice, covers some of the grasses near the stream. What forms of water do you see every day?

TECHTREK
myNGconnect.com

Student eEdition   Vocabulary Games   Digital Library   Enrichment Activities

Fog gathers above this creek in California. The water in the creek is much warmer than the snow-covered banks.

222   223

In Chapter 7, you will learn:

FLORIDA NEXT GENERATION SUNSHINE STATE STANDARDS

SC.3.P.8.1 Measure and compare temperatures of various samples of solids and liquids. **STATES OF WATER, WATER CHANGES STATE, MEASURE TEMPERATURE**

SC.3.P.9.1 Describe the changes water undergoes when it changes state through heating and cooling by using familiar scientific terms such as melting, freezing, boiling, evaporation, and condensation. **STATES OF WATER, WATER CHANGES STATE**

SC.3.P.8.1 Science in a Snap! Measure and compare temperatures of various samples of solids and liquids.

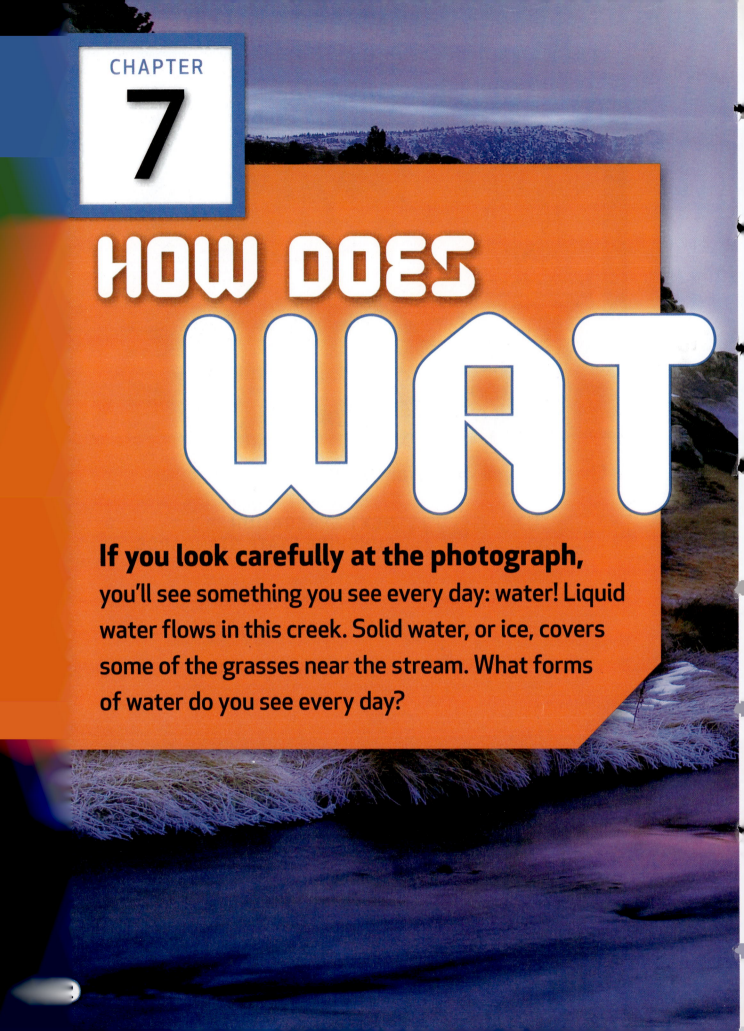

HOW DOES
WAT

If you look carefully at the photograph, you'll see something you see every day: water! Liquid water flows in this creek. Solid water, or ice, covers some of the grasses near the stream. What forms of water do you see every day?

TECH TREK
myNGconnect.com

Student
eEdition

Vocabulary
Games

Digital
Library

Enrichment
Activities

ER CHANGE?

Fog gathers above this creek in California. The water in the creek is much warmer than the snow-covered banks.

SCIENCE VOCABULARY

states of matter
(STĀTS UV MA-ter)

States of matter are the forms in which a material can exist. (p. 226)

Solid ice is one state of matter of water.

solid (SO-lid)

A **solid** is matter that keeps its own shape. (p. 227)

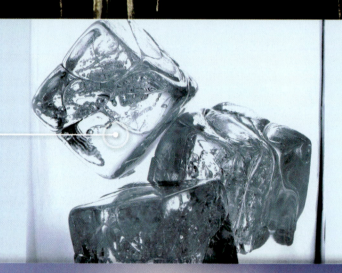

This ice cube is a solid.

liquid (LI-kwid)

A **liquid** is matter that takes the shape of its container. (p. 228)

Liquid water takes the shape of the glass and the vase.

my
Science Vocabulary

condensation
(kon-din-SĀ-shun)

evaporation
(ē-va-pōr-Ā-shun)

gas
(GAS)

liquid
(LI-kwid)

solid
(SO-lid)

states of matter
(STĀTS UV MA-ter)

TECHTREK
myNGconnect.com

Vocabulary
Games

gas (GAS)

A **gas** is matter that spreads to fill a space. (p. 229)

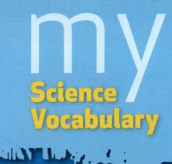

The humidifier adds the gas form of water to the air.

evaporation
(ē-va-pōr-Ā-shun)

Evaporation is the change from a liquid to a gas. (p. 232)

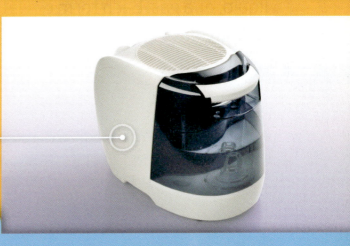

Evaporation causes the water in a puddle to disappear from the road.

condensation
(kon-din-SĀ-shun)

Condensation is the change from a gas to a liquid. (p. 234)

The condensation of water vapor in the air formed the dew on these petals.

States of Water

Water is almost everywhere on Earth! Look for water in this photo of people fishing. Based on your observations, what are some different forms of water?

Sometimes, water flows. When it's frozen, it's hard. There is even water in the air you cannot see. Water can be a solid, liquid, or a gas. Each form of water is called a state. **States of matter** are the forms in which a material can exist.

The water in this stream is always warm. People can fish there during the cold winter, because the water does not freeze.

Water as a Solid The solid form of water is ice. What makes a solid a solid? A solid is matter that keeps its shape. If you take ice out of an ice cube tray and put it in a glass, it still keeps its shape as long as it is cold.

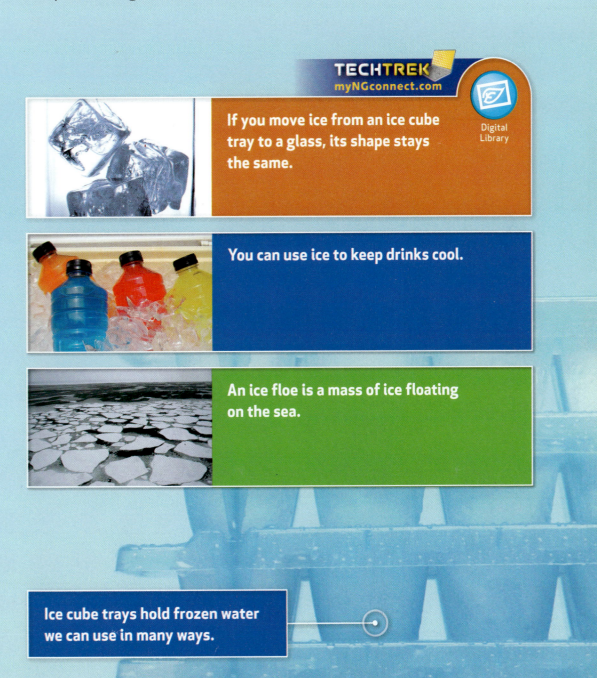

TECHTREK
myNGconnect.com

Digital Library

If you move ice from an ice cube tray to a glass, its shape stays the same.

You can use ice to keep drinks cool.

An ice floe is a mass of ice floating on the sea.

Ice cube trays hold frozen water we can use in many ways.

Water as a Liquid You know what happens when you knock over a glass of water. It forms a puddle. The water in the puddle has a different shape than the water in the glass had. That is because water is a <mark>liquid</mark> . A liquid is matter that takes the shape of its container. That liquid may change shape, but the amount of that liquid is the same no matter what container the liquid is in.

TECHTREK
myNGconnect.com

Enrichment
Activities

Much of Earth is covered with liquid water.

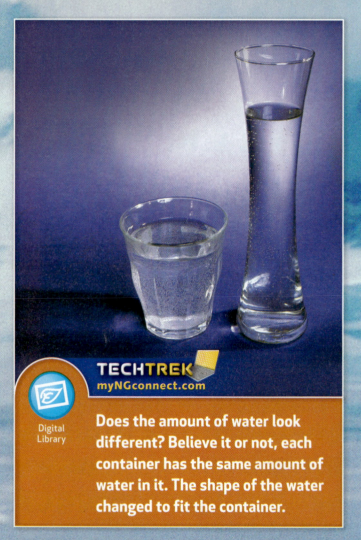

TECHTREK
myNGconnect.com

Digital
Library

Does the amount of water look different? Believe it or not, each container has the same amount of water in it. The shape of the water changed to fit the container.

Water as a Gas Have you ever put a pan of water on the stove to heat the water? You may have noticed that when the water gets really hot, bubbles form in the water. Those bubbles show that a **gas** is forming. A gas is matter that spreads to fill a space. The gas form of water is water vapor. Water vapor is invisible.

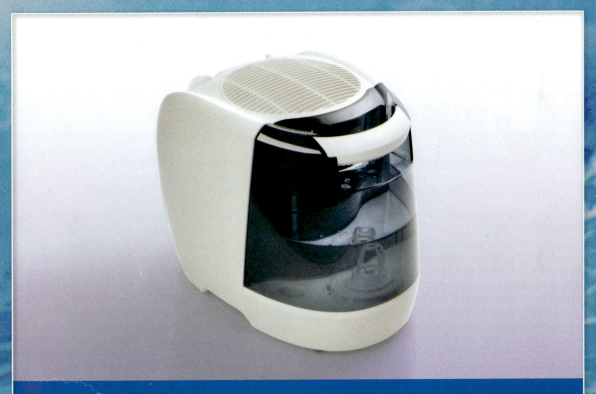

The humidifier adds water vapor to the air so that the air is not as dry.

Before You Move On

1. In what three states can you find water?
2. What is the name of the solid form of water?
3. **Infer**   A state of water is invisible. You can only see it when it enters cooler air. In which state is the water?

Water Changes State

Walking on water is impossible, right? If you have ever skated on ice, though, you have almost done just that! What state is the water in the photo of the ice skaters?

You can find water in all three states in your kitchen. You can even change it from one state to another. Changes of state can happen when the temperature changes.

Skaters enjoy a cool day on an outdoor ice rink in London, England.

Freezing and Melting Ice forms when water freezes. Freezing is the change from a liquid to a solid. Freezing happens at low temperatures. It happens when water cools to its freezing point, which is 0°C (32°F).

Melting is the opposite of freezing. Melting is the change from a solid to a liquid. When the temperature rises above 0°C, ice changes to water.

TECHTREK
myNGconnect.com

Digital Library

When the temperature rises to 0°C and climbs higher, icicles drip, get smaller, and soon disappear.

Evaporation On a humid day, the air feels wet. You cannot see it, but there is water in the air. Water that is a gas is called water vapor. It gets in the air through evaporation. <mark>Evaporation</mark> is the process in which water changes from a liquid to a gas. Evaporation can take place at many temperatures. It takes heat energy to cause evaporation, though.

The liquid water in the puddle evaporated into the air. The gas form of air, called water vapor, is invisible.

Boiling Look at the pot of water boiling on the stove. What do you observe? You probably notice the bubbles that show that water is changing from a liquid to a gas. Water makes this change from a liquid to a gas at 100°C (212°F). That temperature is the boiling point of water. Other liquids have different boiling points.

100°C (212°F) ⟶

Condensation See the white fluffy clouds in the photograph? How did they form? Clouds form because of water. Air contains invisible water in the form of a gas called water vapor. The air high in the sky is cool. As the air cools, tiny drops of water form. Condensation is the change from a gas to a liquid. When water condenses high in the air, it sticks to pieces of dust and floats in the air. When those pieces of dust covered with water come together, they make a cloud you can see.

Clouds form when the temperature of the air changes.

Condensation doesn't happen only high in the air. On the ground, dew can form on grass or flowers. Dew forms when water vapor in the air condenses. If you have a glass of cold lemonade on a warm day, you may notice drops of water on the glass. The drops of water are condensation. The warmer air causes the water vapor in the air to condense on the cold drinking glass.

Dew drops formed on these flower petals. They form when water vapor in the air cools.

Before You Move On

1. Name two actions that cause matter to change from liquid to gas.
2. What changes of state can happen when temperatures get colder?
3. **Apply** After a rainstorm, the sidewalk dries out. What happened to cause this change? Where does the water go?

Measure Temperature

You already know that water changes state when the temperature changes. But how can you measure the temperature? You can use a special tool, a thermometer, to measure the temperature.

Look at the thermometer floating in the pool. The thermometer measures the temperature of liquid water. You could read the thermometer to find out if the water is warm enough for swimming.

A thermometer measures the temperature of this bubbly liquid cooking on the stove.

This floating thermometer measures the temperature of the liquid water in the pool.

You can measure the temperature of solids too. Meat is a solid. Cooks use thermometers to help them decide when meat is finished cooking. When the meat reaches the correct temperature, the cook knows that it is safe for people to eat.

Science in a Snap! Find the Temperature

Look at the oil, corn syrup, and water. Describe them. How are they the same? How are they different?

Put a thermometer in each cup.

What is the temperature of each liquid?

Before You Move On

1. Describe how you would measure the temperature of the soil outside your school.
2. Compare the temperature of water in its liquid state with the temperature of water in its solid state.
3. **Infer** An ice cube falls on the ground outside. A thermometer shows that the outside air temperature is below 0°C. What will happen to the ice cube?

FOG FENCES
IN CENTRAL AMERICA

In some places in Guatemala, it is hard to get fresh water. It does not rain very often. There are few wells. It is hard to build pipes to bring in water from cities that are far away.

The fog fence collects water that helps this girl and her family drink, cook, bathe, and grow food.

The people who live in Guatemala still need water. They need it to drink and cook. They need it to feed their animals and grow plants for food. One way to get water is to build a fog fence. A fog fence uses condensation to collect water.

Fog is a cloud of water droplets floating in the air near the ground. A fog fence is a large plastic net. When wind blows fog through the net, the net gets wet. The water rolls or trickles down into pipes. A net that is 8 meters (26.2 feet) high and 50 meters (164 ft) long collects about 200 liters (52.8 gallons) each day. That amount would fill about four bathtubs.

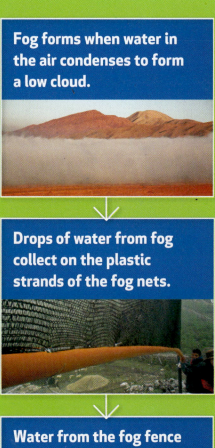

Fog forms when water in the air condenses to form a low cloud.

Drops of water from fog collect on the plastic strands of the fog nets.

Water from the fog fence collects in a large tank.

Water exists in three states: solid, liquid, and gas. Water changes state because of changes in temperature. Freezing and condensation happen when water is cooled. Melting, boiling, and evaporation happen when water is heated.

Big Idea The changes of state in water are caused by changes in temperature.

Solid

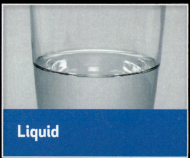

Liquid

Gas

Vocabulary Review

Match the following terms with the correct definition.

A. condensation

B. evaporation

C. gas

D. states
   of matter

E. liquid

F. solid

1. Matter that keeps its own shape

2. Matter that spreads to fill a space

3. Matter that takes the shape of
   its container

4. The change from a gas to a liquid

5. The change from a liquid to a gas

6. The forms in which a material
   can exist

Big Idea Review

1. **Define** Give definitions for these terms: *solid, liquid, gas.* Tell what water is like in each of these states.

2. **Explain** What happens to the shape of a liquid when it is poured into a container with a different shape?

3. **Describe** How is evaporation part of boiling?

4. **Cause and Effect** What would happen to a jar of water vapor if you put it in the refrigerator?

5. **Evaluate** A friend says to you, "Water is very useful in its liquid and solid states, but not in its gas state." How would you respond?

6. **Infer** What kinds of places can use fog fences?

Write About Matter

Describe What is happening in this photograph? What states of matter are there? What change is taking place? How is temperature important?

PHYSICAL SCIENCE EXPERT: ICE SCULPTOR

What Does an Ice Sculptor Do?

Some artists make sculptures out of stone. Others use ice! Steve Brice wins contests for carving beautiful shapes out of solid ice. He works in Chena Hot Springs in the northern state of Alaska. The temperature is below freezing much of the year there. So, his sculptures last for months outside without melting.

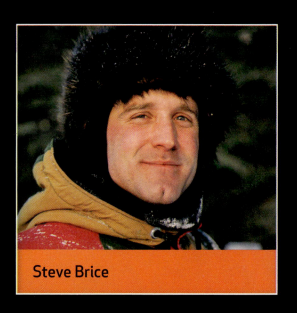

Steve Brice

Colorful lights brighten up the sculptures in Steve Brice's gallery.

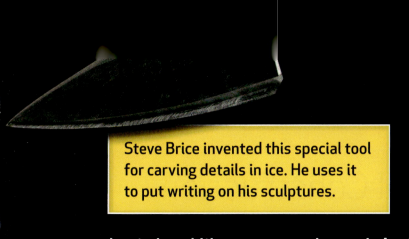

Steve Brice invented this special tool for carving details in ice. He uses it to put writing on his sculptures.

Ice is hard like stone and wood. An ice sculptor uses many of the same tools to carve ice that other sculptors use to shape other solids. Brice uses chain saws, sanders, knives, and drills to shape the ice.

As long as it stays cold, the finished sculptures keep their shape. Every year, the sculptures melt in the spring. Brice makes new ones when it freezes again in winter.

TECHTREK
myNGconnect.com

Digital
Library

Steve Brice uses a special tool to smooth the face of one of his ice sculptures.

BECOME AN EXPERT

Sweden: Ice Hotel Construction

Buildings have to keep their shapes. They cannot bend or flow. This is why they have to be made out of **solid** things. Did you know that a building can be made out of water? A hotel in the far northern part of the country of Sweden is made out of water in its solid state—ice!

> The temperature has to be well below 0°C (32°F) to keep the ice—the hotel—solidly frozen.

solid

A **solid** is matter that keeps its own shape.

Freezing Ice Blocks

In some places, it gets very cold in the winter. When it gets below 0°C (32°F), water freezes. In cold places such as northern Sweden, the **liquid** water in lakes and rivers can freeze. When a river freezes, the water at the top turns solid. The people who build the ice hotel use large blocks of this frozen river water.

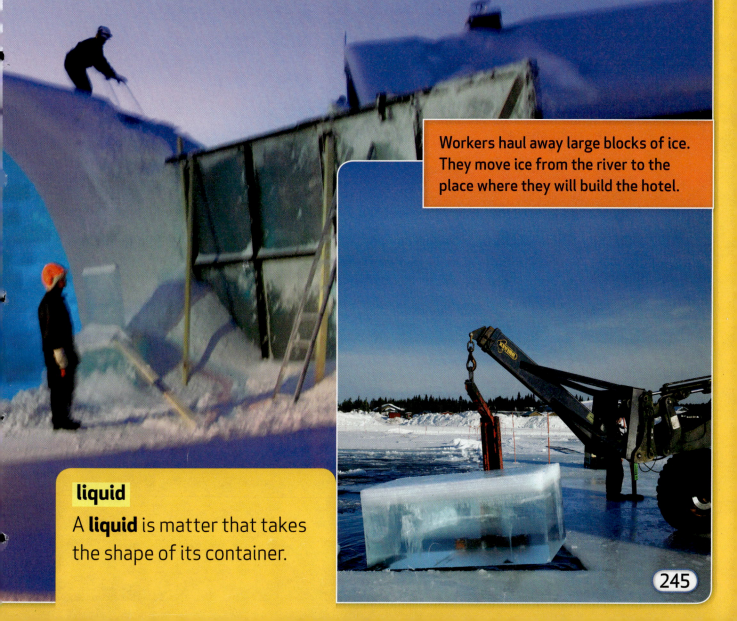

Workers haul away large blocks of ice. They move ice from the river to the place where they will build the hotel.

liquid

A **liquid** is matter that takes the shape of its container.

Building with Ice and Snow

Some parts of the ice hotel are made with packed snow. Snow is a powdery solid. When it is packed down, it makes a hard shape. Builders of the ice hotel use packed snow to make the walls and ceilings of the ice hotel.

1. Forms are set into place. These forms give the snow the right shape.

2. A snow cannon shoots snow onto the forms. It is packed into a solid shape on the form.

3. When the packed snow has the right shape, the form is taken out. It can be used again to build another part of the hotel.

HOW DOES BUILDING WITH ICE WORK?

Winter temperatures drop, and the river freezes.

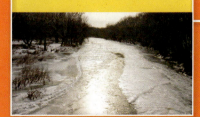

Blocks of ice are cut from the frozen river.

The blocks are shaped and stacked to make pillars, beds, and so on.

Living in Ice

To keep the ice frozen, it must stay cold inside the hotel. The rooms are usually five degrees colder than 0°C. Outside, it can be 20 or 30 degrees colder than that temperature.

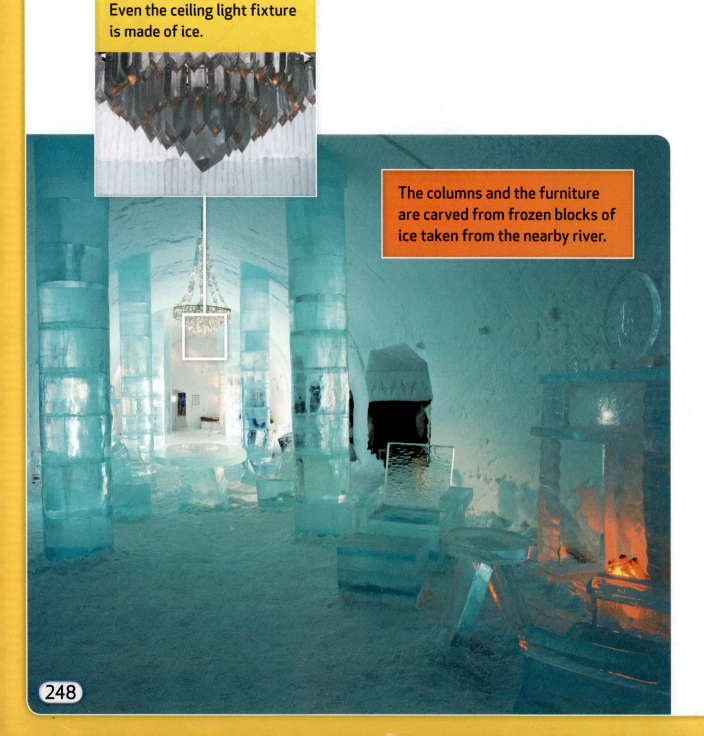

Even the ceiling light fixture is made of ice.

The columns and the furniture are carved from frozen blocks of ice taken from the nearby river.

Seeing Your Breath A guest at the ice hotel might be able to see her breath. When it is cold, sometimes it looks like clouds are coming out of your mouth when you breathe. The puffs are not a **gas** . They are tiny drops of water. They formed when warm air that you breathed out cooled. In other words, **condensation** occurred.

TECHTREK
myNGconnect.com

Digital Library

Guests sleep on beds made of ice and covered with furs.

It is cold enough inside that a guest must wear her jacket.

gas

A **gas** is matter that spreads to fill a space.

condensation

Condensation is the change from a gas to a liquid.

The Hotel Melts

When spring comes, it gets warm again. Soon, the temperature gets above 0°C. The ice of the hotel cannot stay solid. It begins to melt. As the blocks melt, liquid water forms. It drips from the ceiling. It collects in puddles. The hotel is completely gone!

WHERE DID THE ICE HOTEL GO?

Water in nature can change between all three states of matter.

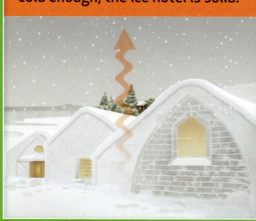

When the temperature outside is cold enough, the ice hotel is solid.

As the ice hotel melts, the gas form of water enters the air.

Gas Again In just two months, all of the ice that made up the hotel melts. The water from the melting ice runs into the river. **Evaporation** occurs, and some water becomes water vapor. The ice hotel's water has changed into all three **states of matter**. Next winter, the water in the river will freeze again. It can be used to build a whole new ice hotel.

When the ice melts, the hotel slowly disappears.

evaporation
Evaporation is the change from a liquid to gas.

states of matter
States of matter are the forms in which a material can exist.

CHAPTER 7

SHARE AND COMPARE

Turn and Talk How do states of matter affect the Ice Hotel? Form a complete answer to this question together with a partner.

Read Select two pages in this section. Practice reading the pages. Then read them aloud to a partner. Talk about why the pages are interesting.

 **Write** Write a conclusion that tells the important ideas you learned about constructing an ice hotel. State what you think is the Big Idea of this section. Share what you wrote with a classmate. Did your classmate make the connection between the states of water and how they are important to constructing the hotel?

 **Draw** Draw a picture of a stage in the making or the melting of the ice hotel. Label the states of matter and the changes of state that are happening. Combine your drawings with those of your classmates to create a guidebook for the Ice Hotel.

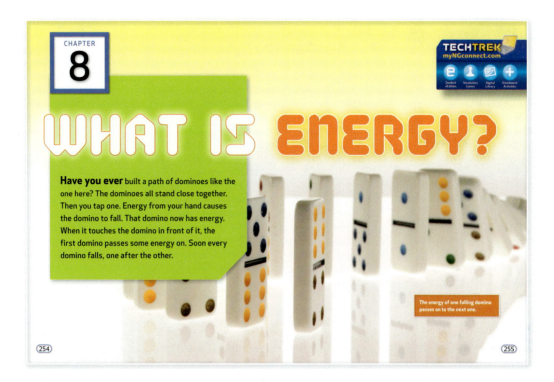

CHAPTER 8

WHAT IS ENERGY?

Have you ever built a path of dominoes like the one here? The dominoes all stand close together. Then you tap one. Energy from your hand causes the domino to fall. That domino now has energy. When it touches the domino in front of it, the first domino passes some energy on. Soon every domino falls, one after the other.

TECHTREK
myNGconnect.com

The energy of one falling domino passes on to the next one.

(254)   (255)

In Chapter 8, you will learn:

FLORIDA NEXT GENERATION SUNSHINE STATE STANDARDS

SC.3.P.10.1 Identify some basic forms of energy such as light, heat, sound, electrical, and mechanical. **ENERGY, MECHANICAL ENERGY, SOUND, ELECTRICAL ENERGY, HEAT**

SC.3.P.10.2 Recognize that energy has the ability to cause motion or create change. **ENERGY, MECHANICAL ENERGY**

SC.3.P.11.2 Investigate, observe, and explain that heat is produced when one object rubs against another, such as rubbing one's hands together. **HEAT**

SC.3.P.11.2 Science in a Snap! Investigate, observe, and explain that heat is produced when one object rubs against another, such as rubbing one's hands together.

WHAT IS

Have you ever built a path of dominoes like the one here? The dominoes all stand close together. Then you tap one. Energy from your hand causes the domino to fall. That domino now has energy. When it touches the domino in front of it, the first domino passes some energy on. Soon every domino falls, one after the other.

ENERGY?

The energy of one falling domino passes on to the next one.

SCIENCE VOCABULARY

energy (EN-ur-jē)

Energy is the ability to do work or cause a change. (p. 258)

These children use energy to play on a playground.

mechanical energy
(mi-CAN-i-kul E-nur-jē)

The **mechanical energy** of an object is its stored energy plus its energy of motion. (p. 261)

This skier has mechanical energy.

sound (SOWND)

Sound is energy that can be heard. (p. 264)

This musician creates sound as he beats a drum.

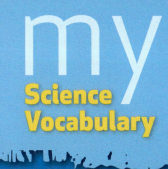
my
Science Vocabulary

electricity
(Ē-lek-TRIS-it-ē)

energy
(EN-ur-jē)

heat
(HĒT)

mechanical energy
(mi-CAN-i-kul E-nur-jē)

sound
(SOWND)

TECHTREK
myNGconnect.com

Vocabulary
Games

electricity (Ē-lek-TRIS-it-ē)

Electricity is energy that flows through wires. (p. 270)

This light bulb uses electricity to light up a room.

heat (HĒT)

Heat is the flow of energy from a warmer object to a cooler object. (p. 274)

Heat flows from the warm soup to the cool spoon.

Energy

Look at the children on the swings. They are doing work. When scientists talk about work, they use this word differently than you might. In science, work means using force to make something move. These children use force to pump back their legs back and forth. They need **energy** to do this. Energy is the ability to do work or cause a change.

TECH**TREK**
myNGconnect.com

Digital Library

The children are doing work as they swing on the playground. They are using energy to cause a change.

In bowling, you give energy to a ball by rolling it down the lane. The ball hits the pins and knocks them down. Energy has gone all the way from you, to the ball, and down the lane to the pins.

Before You Move On

1. Describe energy.
2. Explain how energy and work are related.
3. **Apply** Describe the flow of energy when you are throwing a baseball to your friend.

Mechanical Energy

Look at the picture below. The children ready to sled down the hill have a lot of stored energy. Stored energy is energy that is ready to be used.

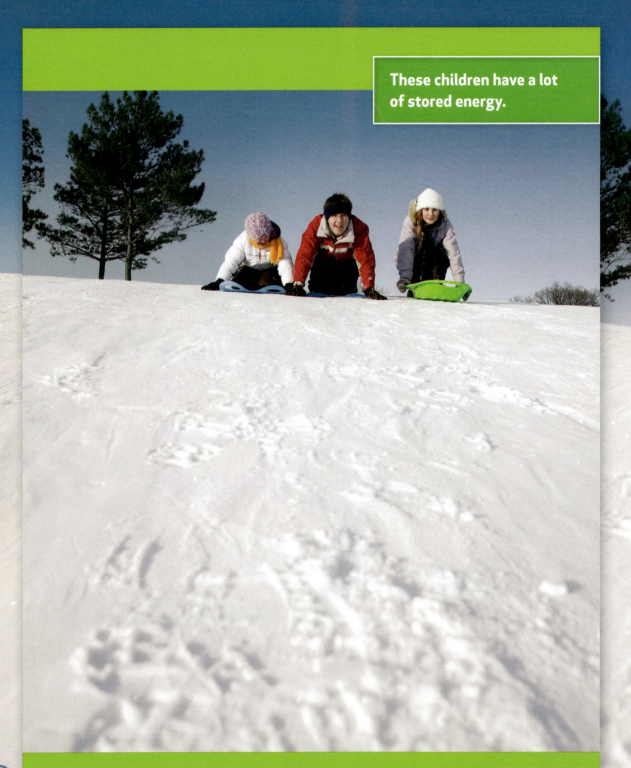

These children have a lot of stored energy.

Look at the children sledding down the hill. Their stored energy has turned into a new type of energy, the energy of motion. All moving things have the energy of motion. The **mechanical energy** of each child is their stored energy plus their energy of motion.

The children have the energy of motion as they sled down the hill.

Let's look at another example of mechanical energy. A skier starts at the top of a mountain. She has stored energy. She is not moving. She does not have any energy of motion.

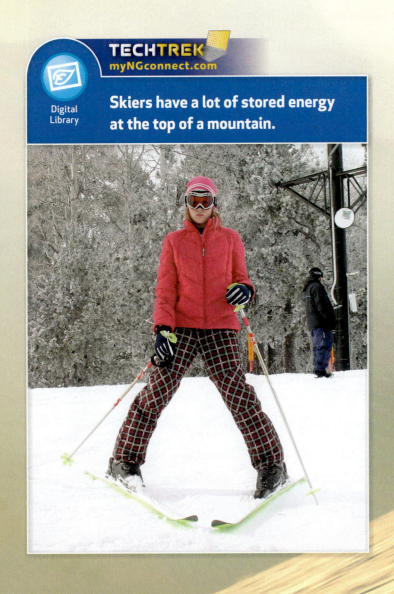

Skiers have a lot of stored energy at the top of a mountain.

Look at this skier. He is in motion. His stored energy changes to energy of motion as he moves down the mountain. The skier's stored energy plus his energy of motion is his mechanical energy.

As the skier moves down the hill, his stored energy changes to energy of motion.

Before You Move On

1. What is mechanical energy?
2. How is stored energy different than energy of motion?
3. **Apply** You are running down a hill. How does your energy change?

Sound

Have you heard someone play a drum? When the stick hits the drum, the drum vibrates, or moves quickly back and forth. When the top of the drum vibrates, the air near the drum begins to vibrate too. These vibrations make sound. **Sound** is energy that can be heard!

This boy creates sound by hitting the skin of a drum.

You hear sound when sound travels to your ear. Look at your ear in a mirror. It is made to collect sound vibrations. The vibrations move to your ear, and you hear the sound.

Volume A jet engine is very loud when you are near it. It has a loud volume. Volume is how loud or soft a sound is. Strong vibrations make loud sounds. People working on the ground at an airport wear special headphones to protect their ears.

Without the headphones, the sound could damage this airport worker's hearing.

When you speak, the vocal chords inside your throat vibrate. When you shout, your vocal chords vibrate more. This makes your voice sound louder.

When you speak quietly, your vocal chords vibrate less. This makes your voice sound softer.

Which boy's vocal cords are vibrating more?

Pitch Look at the animals on this page. What kind of sounds do you think they make?

The kitten has thin, short vocal chords that vibrate quickly. They make a high sound. The tiger has thick, long vocal chords that vibrate more slowly. They make a low sound. Pitch is how high or low a sound is.

The kitten has a high-pitched meow.

The tiger has a low-pitched roar.

Look at the guitar. When the strings move, they vibrate. They make sound. A thick string vibrates slowly. The thick strings make low, deep sounds. This is a low pitch. A thin string can vibrate much faster. The thin strings make high sounds. This is a high pitch.

Thick strings make low-pitched sounds.
Thin strings make high-pitched sounds.

+
Enrichment
Activities

Before You Move On

1. What is sound?

2. Compare volume and pitch.

3. Infer Give an example of a sound with a loud volume and low pitch.

Electrical Energy

A DVD player needs energy to play movies. It uses **electricity**. Electricity is energy that flows through wires. You can watch the movies and hear the sound because of electricity.

The headphones change the electrical energy into sound you can hear.

Some DVD players plug into an outlet in the wall. You can also find DVD players that run on batteries. Batteries have stored energy. The DVD player changes the stored energy to electricity.

All of these things need electricity so they can work.

radio

cell phone

DVD player

DVD

This vacuum gets its power from electricity. Electricity is flowing through the wires inside the vacuum. The electricity becomes energy of motion. The wheels on the vacuum turn. The vacuum takes in air and dirt.

This vacuum cleaner uses electricity.

Electricity also flows through wires to make light bulbs shine. The light bulb turns the electricity into light. The light allows you to see all the objects around you.

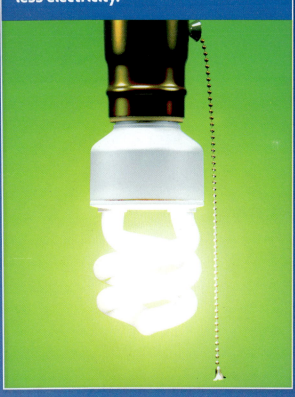

Energy-saving light bulbs use less electricity.

Before You Move On

1. What is electricity?
2. How do electronic devices like DVD players get electricity?
3. **Apply** Choose something electric from your home. How does it change electricity into another form of energy?

Heat

This soup was just cooked. It is very hot. The metal spoon was in a drawer. It is much cooler than the soup. When the spoon is used to stir the soup, it quickly becomes hot. **Heat** moves from the soup to the spoon. Heat is the flow of energy from a warmer object to a cooler object.

The cool spoon heats up when you put it in hot soup.

Heat moves easily though some materials such as metal. The cookies bake on the hot metal cookie sheet. The woman wears cloth oven mitts to take the cookie sheet out of the oven. Heat does not move easily through cloth or plastic.

Thick, cloth oven mitts protect this woman's hand from the heat.

Look at the photo on this page. What do you think is happening to the rocks?

Two rocks are being rubbed together. They get so hot that a spark forms. Heat is made whenever two things rub together.

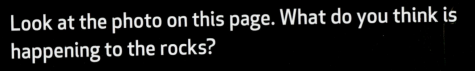

When these two rocks hit each other, they heat up. The rocks may create enough heat to start a fire.

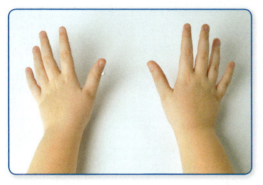

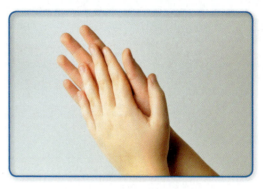

Describe the temperature of your hands.

Rub your hands together for 30 seconds. Rub them fast!

What happened to the temperature of your hands?

Before You Move On

1. What is heat?
2. Why do you use a cloth oven mitt to take a hot tray out of the oven?
3. **Predict** What would happen if you rubbed two ice cubes together?

NATIONAL GEOGRAPHIC

WIND POWER
ENERGY IN THE AIR

Many people today are talking about wind power. But wind power is not new.

Five thousand years ago, the Egyptians found that wind could help them travel. They put sails on their ships to catch the wind. The wind pushed on the sails. Boats traveled on the Nile River more quickly than they ever had.

Wind gives humans energy to move ships and grind grain.

Wind power can do many things. People have used wind power to pull water out of the ground. Ancient windmills were used to crush grain to make flour.

Today windmills change the motion of wind into other kinds of energy such as electricity. The electricity made by windmills is used to power homes, office buildings, and schools. People will continue to use wind power because it is a clean, renewable source of energy.

People are turning back to wind to get energy.

These windmills on the beach of Mykonos, Greece, were once used to crush grain.

Conclusion

Energy is the ability to do work or cause a change. Energy comes in many forms, such as mechanical energy, sound, electricity, and heat.

Big Idea Energy comes in many forms and has the ability to do work or cause a change.

Vocabulary Review

Match the following terms with the correct definition.

A. energy

B. sound

C. heat

D. electricity

E. mechanical energy

1. The flow of energy from a warmer object to a cooler object.

2. The ability to do work or cause a change

3. An object's stored energy plus its energy of motion

4. Energy that can be heard

5. Energy that flows through wires

Big Idea Review

1. Recall What is energy?

2. Describe Use your own words to describe stored energy and energy of motion.

3. Explain Explain why sound is a form of energy.

4. Compare How do a fan and a heater use electricity?

5. Predict What happens when you rub two objects together?

6. Apply Name one way that electricity makes your life easier and explain why.

Write About Energy

Explain Look at the two sets of strings. Which ones make a higher sound? Which ones make a lower sound? Explain why in your own words.

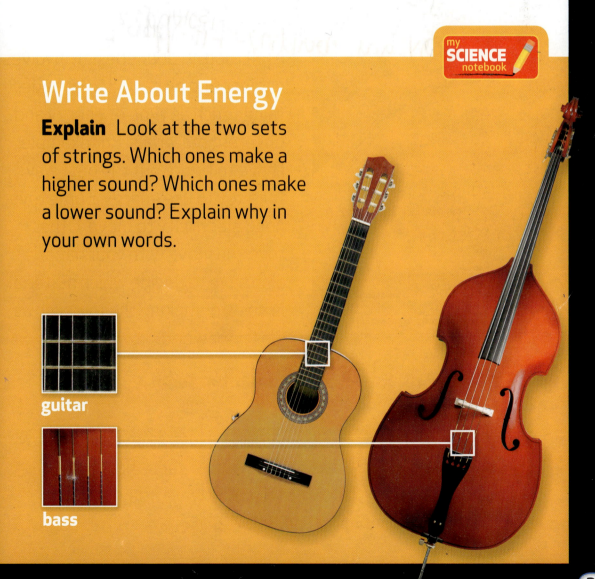

guitar

bass

CHAPTER
8

PHYSICAL SCIENCE EXPERT: PRODUCT DESIGNER

Product designers make all sorts of things that use energy including toys! Judy Lee uses her creativity and knowledge of energy to design new toys for kids of all ages.

Judy Lee working in her workshop.

Q: What is your job?

I'm a product designer. I have designed toys, pet products, and even food. Product design involves engineering by having to consider things such as type of energy and different materials. For example, I have to figure out the best way for a toy to use energy. Some toys need electricity to move. Other toys you move with your hand. I also have to decide which material is best for making the toy, such as plastic or wood.

Q: When did you first know you wanted to be a product designer?

When I was a kid, I loved to build things. I loved figuring out how things worked. Then I would start thinking about how I could make things better. I still think the same way today. I guess that's why I'm a product designer.

TECHTREK
myNGconnect.com

Student
eEdition

Digital
Library

Q: What is a typical day like for you?

Each day is different. When I start a project, I talk to lots of different people. I'll take their ideas and create something, such as a toy. I'll take the toy back to them and hear what they think. Then, I'll use their feedback to make the toy better. I spend a lot of my time talking with people and making things.

Q: What is your favorite thing about being a designer?

I love solving problems. I love building things that use energy in different ways. It's fun to make new things.

Judy Lee builds a new toy.

Judy Lee uses special tools to create her toys.

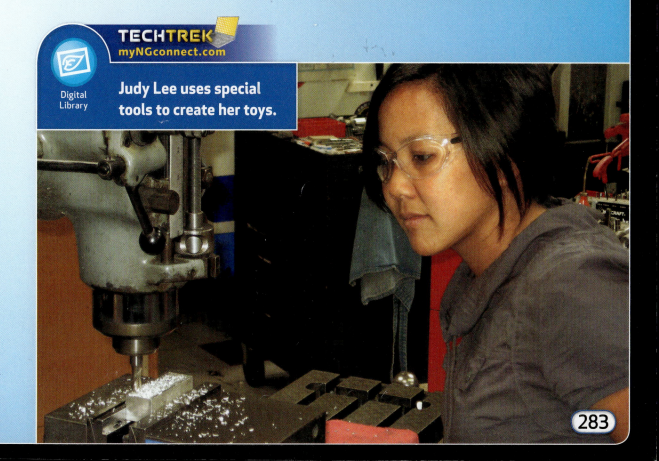

NATIONAL GEOGRAPHIC

BECOME AN EXPERT

Use Some Energy: Have Some Fun!

Toys come in all shapes and sizes. They are fun to use, and they can teach us a lot about **energy** and work.

This pinwheel has energy of motion when you blow on it.

energy
Energy is the ability to do work or cause a change.

284

This robot is walking. It has energy of motion.

Have you ever used a jack-in-the-box? There's a spring inside that is pushed together as you crank the handle. The tighter the spring gets, the more stored energy it has. What happens when you let the spring go? Boing! The clown pops out of the box!

The jack-in-the-box pops out when stored energy changes into energy of motion.

A pogo stick also uses springs. Each time you jump on the stick, the springs push together. The springs gain stored energy.

The pogo stick uses metal springs to bounce up and down.

When the springs go back to their original size the stored energy changes to energy of motion. The energy of motion causes you to bounce in the air. The pogo stick's stored energy plus its energy of motion is its **mechanical energy** .

MECHANICAL ENERGY: POGO STICK

This pogo stick has stored energy because its springs are pushed down.

This pogo stick has changed the stored energy into the energy of motion.

mechanical energy

The **mechanical energy** of an object is its stored energy plus its energy of motion.

287

There are many toys that use **electricity** . Electric train sets are fun to set up. You can hook the tracks together in a design. When you're finished, the train will move along the track. Electricity flows through wires to the tracks. Electricity causes the train to move.

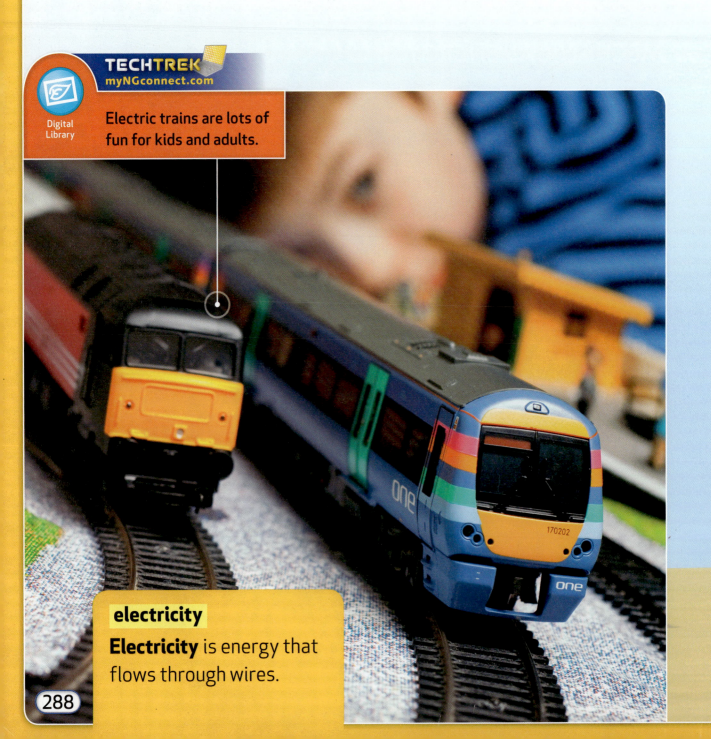

TECHTREK
myNGconnect.com

Digital Library

Electric trains are lots of fun for kids and adults.

electricity

Electricity is energy that flows through wires.

Some toys use **heat**. Heat is the flow of energy from a warmer object to a cooler object. This toy oven heats up and bakes little cakes. The heat from the warm oven flows to the cooler cake batter. The heat from the oven cooks the batter. When you take the cakes out of the oven, the heat flows from the cakes into the air. This is how they cool off. Now they are ready to eat.

Heat from a bulb inside the oven cooks the food.

heat

Heat is the flow of energy from a warmer object to a cooler object.

Have you ever played telephone? You attach two cans to a string. One person talks while the other one listens. **Sound** travels along the string. Sounds are caused by vibrations. When a person talks, the vibrations travel to your ear and make your eardrum vibrate. This is how you hear a sound.

Sounds can have high or low pitches. The girl can change the pitch of her voice while she talks into the telephone.

In a game of telephone, one girl makes a sound with her voice while the other one listens.

sound
Sound is energy that can be heard.

All of these toys use energy. Can you describe how they work? Think about what makes them move or make a sound.

You can use a mallet to make music on this xylophone. The shorter, flat pieces have higher pitches than the longer ones.

This remote-controlled car uses electricity to move.

CHAPTER 8
SHARE AND COMPARE

Turn and Talk How does your favorite toy use energy? Form a complete answer to this question together with a partner.

Read Select two pages in this section. Practice reading the pages. Then read them aloud to a partner. Talk about why the pages are interesting.

Write Write a conclusion that tells the important ideas about energy and toys. State what you think is the Big Idea of this section. Share what you wrote with a classmate. Compare your conclusions.

Draw Draw a picture of your favorite toy. Add labels to your drawing. Share your drawing with a partner. Describe how your toy uses energy to do work.

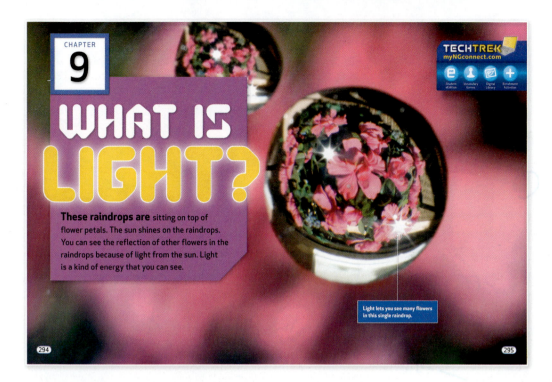

CHAPTER 9

WHAT IS LIGHT?

These raindrops are sitting on top of flower petals. The sun shines on the raindrops. You can see the reflection of other flowers in the raindrops because of light from the sun. Light is a kind of energy that you can see.

TECHTREK
myNGconnect.com

Student eEdition · Vocabulary Games · Digital Library · Enrichment Activities

Light lets you see many flowers in this single raindrop.

294   295

In Chapter 9, you will learn:

FLORIDA NEXT GENERATION SUNSHINE STATE STANDARDS

SC.3.P.10.1 Identify some basic forms of energy such as light, heat, sound, electrical, and mechanical. **SOURCES OF LIGHT, REFLECTION**

SC.3.P.10.3 Demonstrate that light travels in a straight line until it strikes an object or travels from one medium to another. **REFLECTION, REFRACTION, SHADOWS**

SC.3.P.10.4 Demonstrate that light can be reflected, refracted, and absorbed. **REFLECTION, REFRACTION, ABSORPTION**

SC.3.P.11.1 Investigate, observe, and explain that things that give off light often also give off heat. **SOURCES OF LIGHT**

SC.3.P.10.4 **Science in a Snap!** Demonstrate that light can be reflected, refracted, and absorbed.

WHAT IS LIGHT?

These raindrops are sitting on top of flower petals. The sun shines on the raindrops. You can see the reflection of other flowers in the raindrops because of light from the sun. Light is a kind of energy that you can see.

TECHTREK
myNGconnect.com

 Student
eEdition

 Vocabulary
Games

 Digital
Library

 Enrichment
Activities

Light lets you see many flowers in this single raindrop.

SCIENCE VOCABULARY

light (LĪT)

Light is a kind of energy you can see. (p. 298)

Light from the sun shines on a raindrop.

reflection (rē-FLEK-shun)

Reflection is the bouncing of light off of an object. (p. 300)

The reflection in a mirror is an example of how light bounces off a smooth, shiny object.

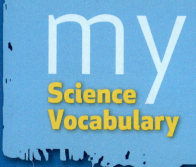

my
Science Vocabulary

absorption
(ab-ZORB-shun)

light
(LĪT)

reflection
(rē-FLEK-shun)

refraction
(rē-FRAK-shun)

TECHTREK
myNGconnect.com

Vocabulary
Games

refraction (rē-FRAK-shun)

Refraction is the bending of light when it moves through one kind of matter to another. (p. 304)

When the light hits the water, the light changes direction. It bends, or refracts.

absorption (ab-ZORB-shun)

Absorption is the taking in of light by a material. (p. 306)

Absorption of light causes objects to heat up.

Where does light come from? The greatest source of light energy on Earth is the sun. But look around you. Light bulbs, fire, fireflies, and the stars are also sources of light.

Living things all around the world need energy from the sun.

Many sources of light also give off heat. When you sit in the sun, you can see the light. You can also feel its heat. You can see and feel light and heat from other light sources, too. Think about what happens when you turn on a lamp. The light bulb in the lamp gives off light. But be careful. Never touch a lit light bulb. It also gives off heat.

Campers sit by a campfire on a cool, dark night. The campfire helps them see what is around them. It also keeps the campers warm.

Light from the sun is natural light. Light from a light bulb is artificial light.

Before You Move On

1. Name some sources of light.
2. Why is the sun important?
3. **Analyze** How is the sun like a light bulb? How is it different?

Reflection

Have you ever seen images in a large body of water? This happens because light can bounce off of the water. A **reflection** is light bouncing off a surface. Light always moves in a straight line from the light source to that surface. Then the light bounces off the surface in another straight path.

Water is a smooth, shiny surface that reflects light. Here, the reflection of the fall trees shows on the shiny water of the lake.

Smooth and shiny surfaces, such as mirrors, reflect most of the light that hits them. A mirror is piece of glass with a metal coating on the back. When light hits the surface, it reflects an image that you can see.

Most of the light that hits the mirror bounces back. It lets you see a clear, sharp reflection.

How People See Objects People need light to see objects. Look at this picture. The girl sees the red apple. She is able to see it because light reflects off the apple and travels in a straight line to her eyes.

The girls can see the apple because light bounces off the surface of the apple.

Science in a Snap! Investigating Reflection

Draw a circle on an index card. Place the index card on your desk and hold the piece of foil above the card with the shiny side facing the card.

Shine the flashlight onto the shiny side of the foil. Try to get the light from the flashlight to reflect into the circle on the card.

What did you have to do to get the light to shine in the circle?

Before You Move On

1. What happens to light when it hits a smooth, shiny surface?
2. Could you see a clear, sharp reflection in a piece of crumpled shiny wrapping paper? Why or why not?
3. **Infer** How does reflected light allow you to read?

Refraction

You've probably used a straw to drink water before. Did you ever notice that the straw can look bent from the side of the glass? As light passes from air to water, it bends, or changes direction. The straw shows that light bent at the place where it passed from air into the water. Light bending as it passes from one kind of matter to another is called **refraction** .

Refraction makes the straw look like it is in two parts.

The picture below shows a prism. A prism is a clear glass or plastic object. It often is shaped like a triangle. What happens to light that passes through a prism? The light bends as it passes through the prism. Each color that makes up the light bends a different amount. This causes light to break into the colors of the rainbow.

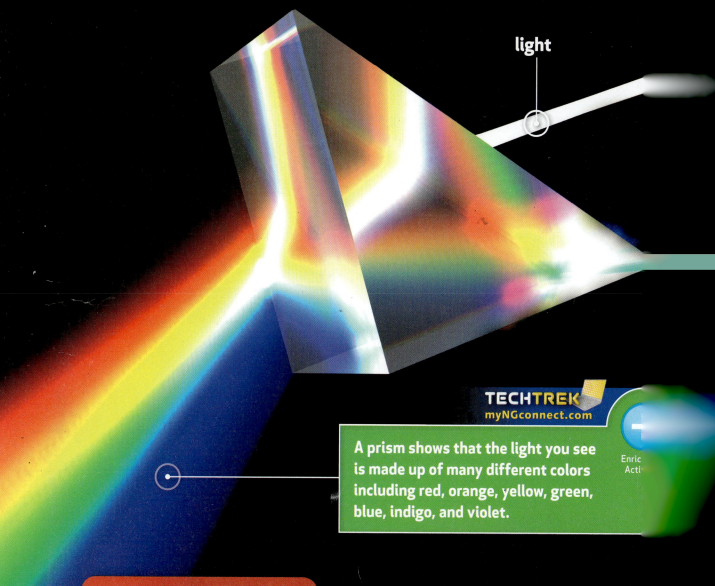

light

A prism shows that the light you see is made up of many different colors including red, orange, yellow, green, blue, indigo, and violet.

Enric
Acti

Before You Move On

1. What happens to light as it passes from air to water?
2. What happens to light when it passes through a prism?
3. Analyze How is refraction different from reflection?

Absorption

Have you ever touched a black car on a hot day? The surface can get really hot. A black object absorbs all of the light that hits it. **Absorption** is the taking in of light by a material. The more light an object absorbs, the faster the object heats up.

The surface of the black car absorbs more light than the surfaces of the other cars.

A white object does not absorb much light. Look at the photos. If you are out in the hot sun, which cap would you wear? If you want to stay cool, you would choose the white cap. The dark cap will get hot faster than the white cap.

The dark-colored cap absorbs more light than the light-colored cap.

Before You Move On

1. What is absorption?
2. Would a dark blue object or light blue object absorb more light?
3. **Evaluate** You are outside in the sun in white clothes. You feel comfortable. Your friend is wearing black. He is complaining about being hot. What is happening?

Shadows

You need three things to make a shadow: a light source, an object, and an area where the light cannot reach.

You can make hand shadows by blocking the light from a flashlight. Since light always travels in a straight line, the light cannot curve around your hand. A shadow forms in the area where the light cannot reach.

TECHTREK
myNGconnect.com

Digital Library

The shadow has formed on the wall because the light could not reach it.

At noon, the sun is high in the sky. Shadows are short. Early in the morning and later in the day, the sun is lower in the sky. Shadows get longer. What can you tell about the time of the day from the shadows of the children?

The long shadows mean the sun is low in the sky. It is early morning or late in the afternoon.

Before You Move On

1. What three things do you need to make a shadow?
2. It's a sunny day. You stand on the sidewalk at noon. You stand in the same place in the early evening. Your shadow is different each time. Why?
3. **Infer** How can you make a shadow disappear?

NATIONAL GEOGRAPHIC

LIGHT POLLUTION
IN THE SKY

As cities grow, their lights are blocking our view of the night sky. We can't see the stars because city lights are so bright. This is called light pollution.

Light pollution harms wildlife. Some animals are active at night. For example, sea turtles lay their eggs at night. However, turtles will not crawl up onto beaches that are brightly lit. As a result, they do not lay their eggs. Some turtle populations are endangered because the turtles don't have dark beaches on which to lay their eggs.

Lights off! See the stars in the night sky.

Lights on! Light reflects off tiny particles in the atmosphere. This causes the glow that blocks the view of the stars in the sky.

Light pollution can be reduced by using lights that are designed to keep light from shining upward and outward. Decreasing light pollution will keep the night sky dark and will benefit many living things.

Sea turtles make nests on beaches at night and lay their eggs.

Conclusion

Light is energy that you can see. Light can be reflected, refracted, or absorbed, depending on what object it hits.

Big Idea Light energy travels in a straight line and comes from sources that give off light and often give off heat.

reflection	refraction	absorption
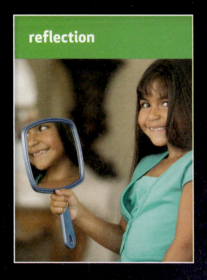	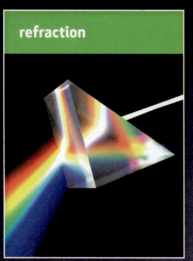	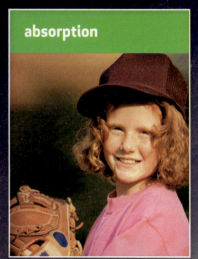

Vocabulary Review

Match the following terms with the correct definition.

A. light   **1.** The bouncing of light off of an object

B. reflection   **2.** The taking in of light by a material

C. refraction   **3.** A kind of energy you can see

D. absorption   **4.** The bending of light when it moves through one kind of matter to another

Big Idea Review

1. Cause and Effect What happens when light hits an object?

2. Recall How do people see objects?

3. Analyze How can a mirror help you see what's behind you?

4. Explain Why does light cause some materials to heat up faster than others?

5. Predict What will happen when an object blocks the path of light?

6. Apply Imagine putting a pencil into a glass of water. Why might the pencil look like it bends?

Write About Refraction

Explain What is happening in this photo? How is the mist from the waterfall acting as a prism?

PHYSICAL SCIENCE EXPERT: INVENTOR

Inventor: Dr. Patricia Bath

Imagine inventing a device that could help people to see better! Dr. Patricia Bath did just that. She discovered that laser light could help.

Dr. Bath is an ophthalmologist. An ophthalmologist is a doctor who does eye surgery and treats diseases of the eye. Dr. Bath invented the laserphaco. This machine is used to remove cataracts. A cataract is a cloudy film over part of the eye. Cataracts can cause blurred vision.

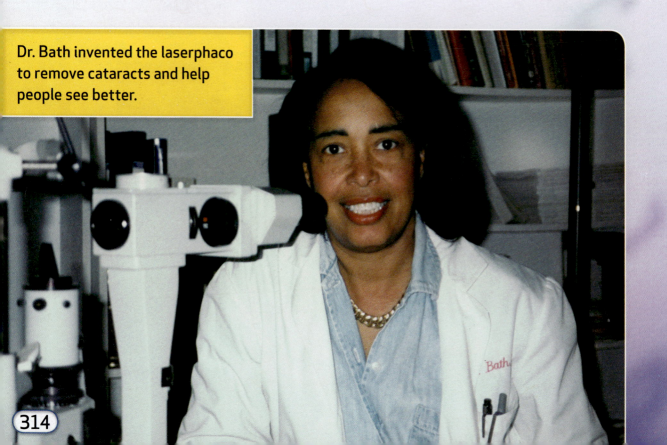

Dr. Bath invented the laserphaco to remove cataracts and help people see better.

TECHTREK
myNGconnect.com

Student
eEdition

Digital
Library

Dr. Bath explained, "When I discovered I could remove cataracts with the laser, I invented a device and called it laserphaco." This is just one of the many ways Dr. Bath has helped people to see.

Dr. Bath has done a lot of other research. She has studied how to use a laser to make surgery methods better. She also works to help prevent blindness.

TECHTREK
myNGconnect.com

Digital
Library

Dr. Bath looks through a microscope as she performs eye surgery. She holds her laserphaco device in her left hand.

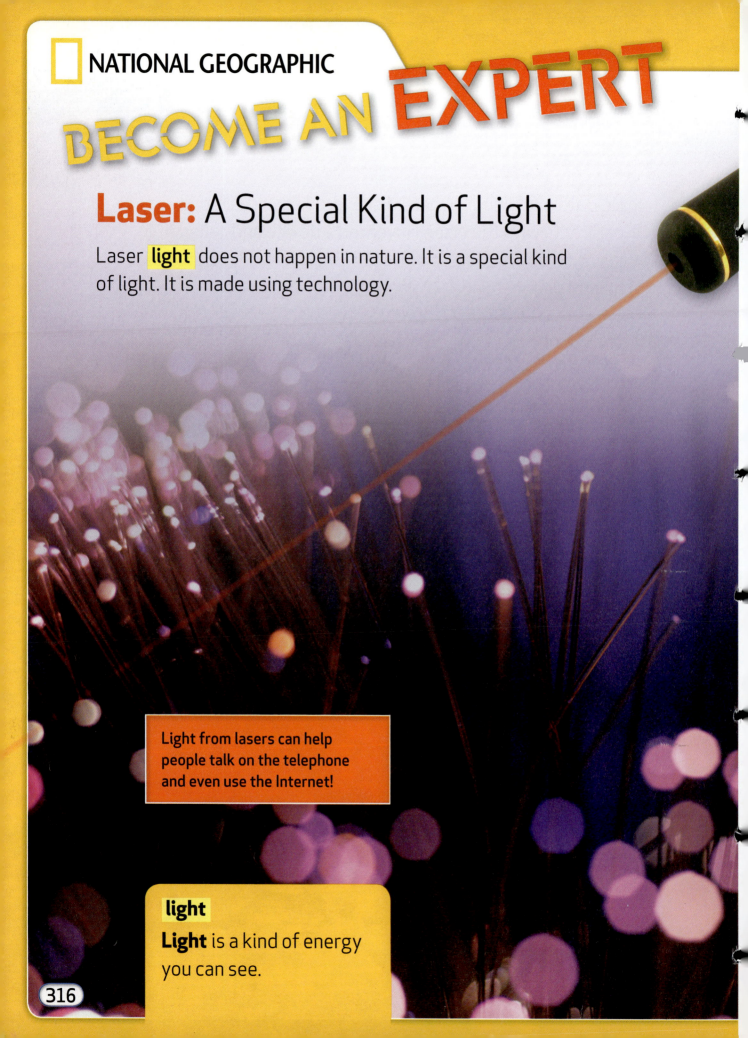

Laser: A Special Kind of Light

Laser **light** does not happen in nature. It is a special kind of light. It is made using technology.

Light from lasers can help people talk on the telephone and even use the Internet!

light

Light is a kind of energy you can see.

What makes laser light so special? It is not like regular light from a bulb. It is not like the natural light from the sun. Laser light does not have different colors of light. It is all one color. Light from the sun and light bulbs goes in all directions. Laser light goes in one direction.

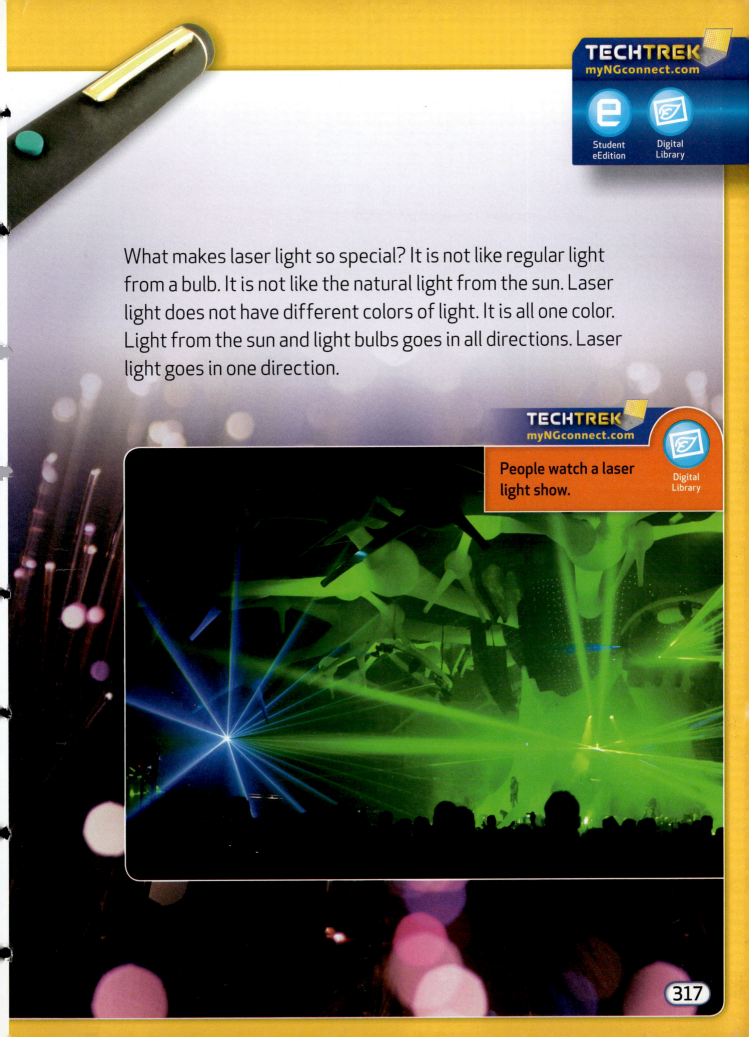

People watch a laser light show.

The First Laser

The first laser was made in 1960. A ruby crystal was used. It was placed between two mirrors. The mirrors helped light reflect back and forth. This made a very powerful laser light. The ruby crystal showed the **reflection** of red light. The result was a ruby-colored laser light.

THE FIRST RUBY LASER

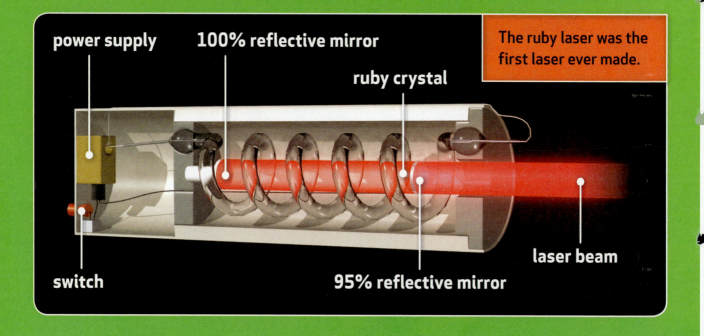

power supply

100% reflective mirror

ruby crystal

The ruby laser was the first laser ever made.

laser beam

switch

95% reflective mirror

reflection

Reflection is the bouncing of light off of an object.

What Makes Lasers Useful?

We have learned a lot about lasers since 1960. We know laser light is powerful. A laser light beam can travel great distances—even to the moon! It is also easy to control. It can be pointed at the exact spot where it is needed. These traits make lasers very useful. Lasers are used in medicine and construction. They are also used in entertainment and in outer space. Lasers are everywhere.

A research scientist uses a laser beam in his work.

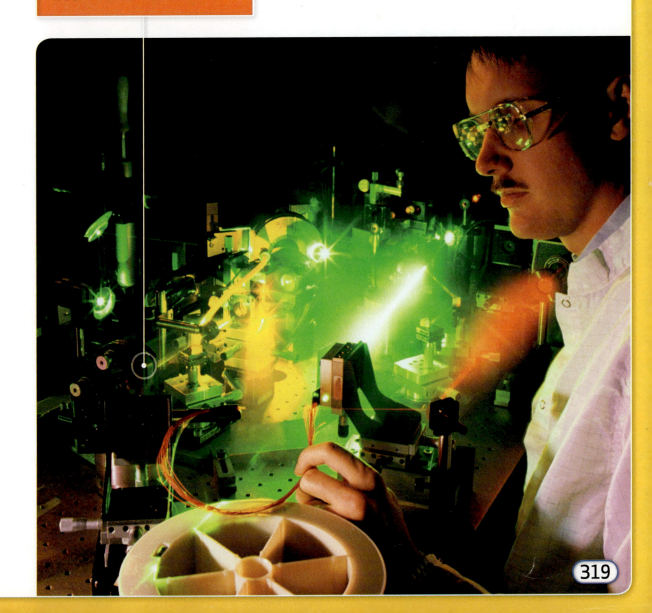

How Eye Doctors Use Lasers

Do you know people who wear glasses? If they take off their glasses, things look out of focus or blurry. This happens because some people's eyes do not refract light correctly. **Refraction** is the bending of light. Eye doctors can use lasers to change the shape of the eye. After laser eye surgery, the eye can correctly refract the light. Then things are in focus and not blurry anymore.

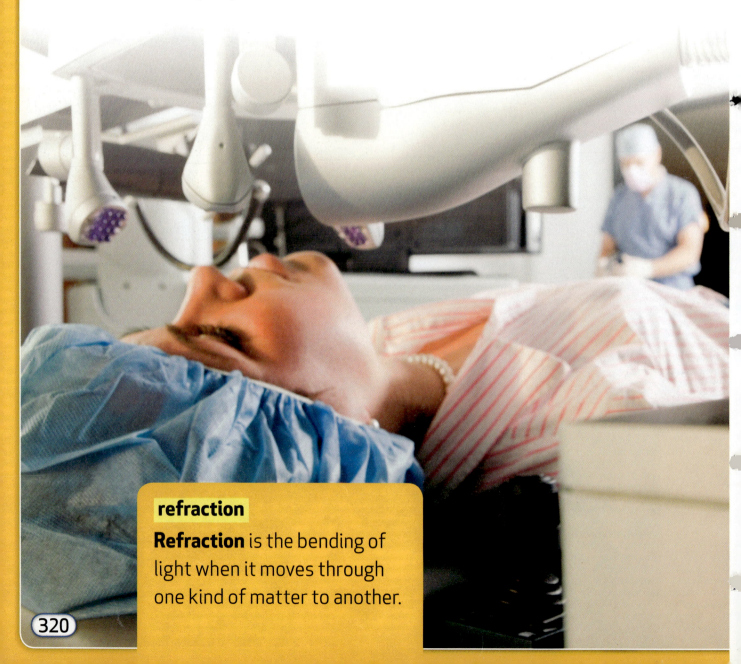

refraction

Refraction is the bending of light when it moves through one kind of matter to another.

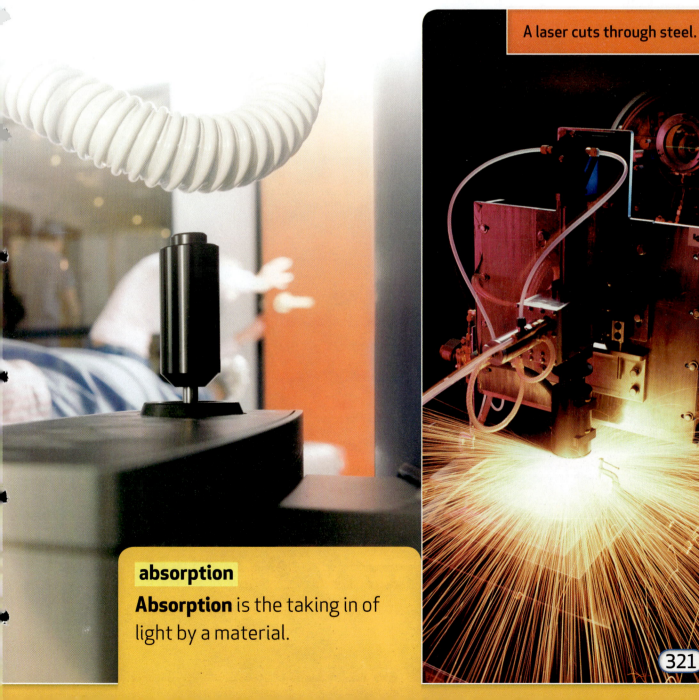

Lasers Shape Metal

A powerful laser can cut through strong metals including steel. The metal **absorbs** , or takes in, the light energy. This makes the metal heat up until it melts. Lasers cut metal precisely. They are sometimes used to engrave small pieces of metal.

A laser cuts through steel.

absorption

Absorption is the taking in of light by a material.

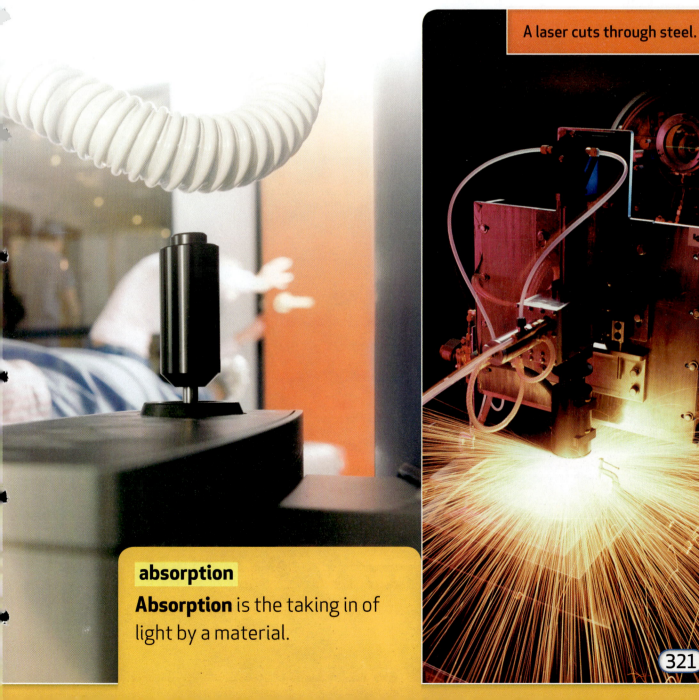

Lasers All Around

Look at the many ways we use lasers in our lives.

Remember, lasers are narrow. They are also powerful. They are easy to control.

These thin cables carry laser light. The light, in turn, delivers signals such as sound, data, and images to people's computers, telephones, and televisions. Many cables can fit through the eye of this needle!

Lasers are used to read bar codes on items in the supermarket.

A special laser device cleans centuries of dirt from stone statues or stone ornaments on buildings without harming the stone underneath.

CHAPTER 9

SHARE AND COMPARE

Turn and Talk How would you compare laser light and sunlight? Form a complete answer to this question together with a partner.

Read Select two pages in this section. Practice reading the pages. Then read them aloud to a partner. Talk about why the pages are interesting.

Write Write a conclusion that summarizes what you learned about lasers. State what you think is the Big Idea of this section. Share what you wrote with a classmate. Compare your conclusions.

Draw Draw a picture of how a laser is used. Include labels. Share your drawing with a classmate. Tell how your drawings are alike and how they are different.

Glossary

A

Absorption (ab-ZORB-shun)
Absorption is the taking in of light by a material. (p. 306)

B

Backbone (BAK-bōn)
A backbone is a string of separate bones that fit together to protect the main nerve cord in some animals. (p. 51)

Brightness (BRĪT-nes)
Brightness is the amount of light that reaches your eye from an object such as a star. (p. 160)

C

Classify (CLA-si-fī)
To classify is to place into groups based on characteristics. (p. 50)

Condensation (kon-din-SĀ-shun)
Condensation is the change from a gas to a liquid. (p. 234)

D

Deciduous (dē-CID-yū-us)
A deciduous plant sheds its leaves every year. (p. 90)

E

Electricity (ē-lek-TRIS-it-ē)
Electricity is energy that flows through wires. (p. 270)

Energy (EN-ur-jē)
Energy is the ability to do work or cause a change. (pp. 128, 258)

Environment (en-VĪ-run-ment)
An environment is all of the living and nonliving things surrounding an organism. (p. 20)

Evaporation (ē-va-pōr-Ā-shun)
Evaporation is the change from a liquid to a gas. (p. 232)

These pine trees live in a windy environment.

Evergreen (EV-ur-grēn)
An evergreen is a plant that keeps its green leaves all year. (p. 91)

G

Gas (GAS)
A gas is matter that spreads to fill a space. (p. 229)

Germinate (JUR-muh-nāt)
A seed germinates when it begins to grow. (p. 21)

Gravity (GRAV-u-tē)
Gravity is a force that pulls objects toward each other. (p. 136)

H

Heat (HĒT)
Heat is the flow of energy from a warmer object to a cooler object. (p. 274)

Heat flows from the warm soup to the cool spoon.

Hibernate (HĪ-bur-nāt)
When animals hibernate, they go into a state that is like a deep sleep during cold winter months. (p. 97)

I

Invertebrate (in-VUR-tuh-brit)
An invertebrate is an animal without a backbone. (p. 53)

This ant is an invertebrate because it does not have a backbone.

L

Light (LĪT)
Light is a kind of energy you can see. (pp. 129, 298)

Liquid (LI-kwid)
A liquid is matter that takes the shape of its container. (p. 228)

M

Mass (MAS)
Mass is the amount of matter in an object. (p. 200)

Matter (MA-ter)
Matter is anything that has mass and takes up space. (p. 194)

Mechanical energy (mi-CAN-i-kul E-nur-jē)
The mechanical energy of an object is its stored energy plus its energy of motion. (p. 261)

Migrate (MĪ-grāt)
When animals migrate they move to another place to meet their basic needs. (p. 98)

O

Organism (OR-guh-niz-uhm)
An organism is a living thing. (p. 10)

P

Pollen (POL-uhn)
Pollen is a powder made by a male cone or the male parts of a flower. (p. 26)

Property (PROP-ur-tē)
A property is something about an object that you can observe. (p. 160)

R

Reflection (rē-FLEK-shun)
Reflection is the bouncing of light off of an object. (p. 300)

Refraction (rē-FRAK-shun)
Refraction is the bending of light when it moves through one kind of matter to another. (p. 304)

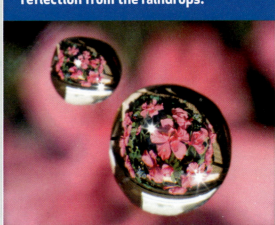

You can see these flowers because of reflection from the raindrops.

The airplane and buildings are made of matter.

Reproduce (rē-pru-DUS)
To reproduce is to make more of its kind. (p. 26)

S

Season (SĒ-zun)
A season is a time of year with certain weather patterns and day lengths. (p. 90)

Solid (SO-lid)
A solid is matter that keeps its own shape. (p. 227)

Sound (SOWND)
Sound is energy that can be heard. (p. 264)

Spore (SPOR)
A spore is a tiny part of a fern or moss that can grow into a new plant. (p. 30)

Star (STAR)
A star is a glowing ball of hot gases. (p. 158)

States of matter (STĀTS UV MA-ter)
States of matter are the forms in which a material can exist. (p. 226)

Sun (SUN)
The sun is the star that is nearest to Earth. (p. 126)

T

Telescope (TEL-uh-scōp)
A telescope is a tool that magnifies objects and makes them look closer and bigger. (p. 166)

Temperature (TEM-pur-ah-chur)
Temperature is a measure of how hot or cold something is. (p. 133)

Texture (TEKS-chur)
Texture describes the surface of any area made up of matter. (p. 198)

Transform (trans-FORM)
To transform is to change. (p. 131)

V

Vertebrate (VUR-tuh-brit)
A vertebrate is an animal with a backbone. (p. 52)

Volume (VOL-yum)
Volume is the amount of space matter takes up. (p. 202)

You can see more stars with a telescope than with just your eyes.

Index

Credits